SciQuest

Milo's Mission

SciQuest: Milo's Mission

First Edition 2026

Email: support@elementalscience.com

ISBN # 978-1-953490-36-0
Printed In USA For worldwide distribution

For more copies write to :
Elemental Science
PO Box 4583
Maryville, TN 37802
support@elementalscience.com

Copyright Policy

Lab Journal At-A-Glance

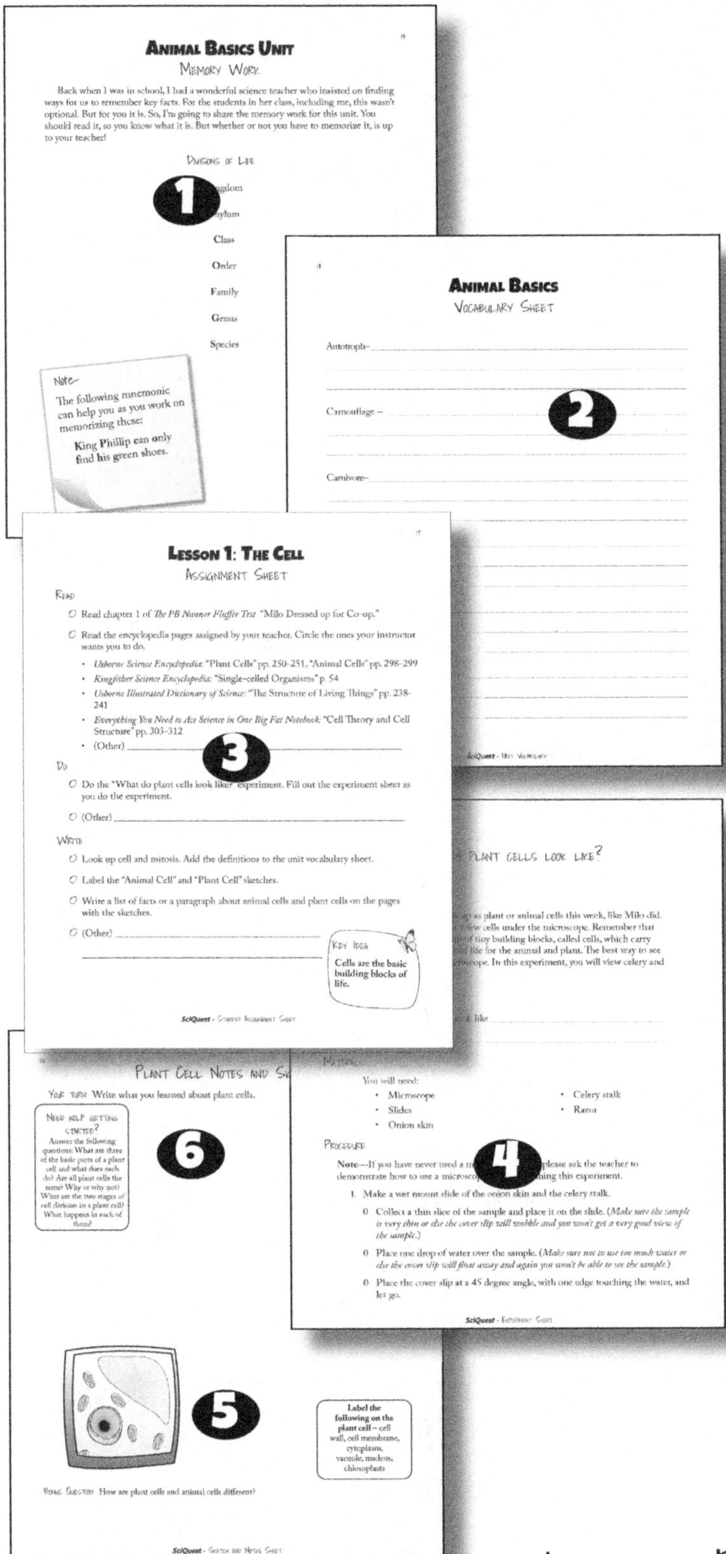

Go on a scientific quest to learn about cells, animals, and our planet.

1. Unit Memory Work

Memorize key facts about the unit's topics with these optional memory work selections.

2. Unit Glossary

Keep all the unit vocabulary in one place. The glossary gives space to write the definitions for each vocabulary word.

3. Student Assignment Sheet

See the plan for the week on the student assignment sheets, broken into reading, doing, and writing sections. This sheet has all the tasks for the week with checkboxes for you to use as you complete the work.

4. Experiment Sheet

Do an experiment related to the topic for the week. The Experiment Sheet contains the introduction and the procedure for the experiment. It also has space to record the hypothesis, materials, observations, results, and conclusion.

5. Pre-drawn Sketch

Color the pre-drawn sketch relating to the week's topic. Then, label it with the information given on the page.

6. Written Assignment

Know the questions for discussion time, and have plenty of space to write either a list of facts, an outline, or a summary.

Additional Resources

The following page contains quick links to the activities suggested in this guide, along with several helpful downloads:

💻 https://elementalscience.com/blogs/resources/sq100

Table of Contents

SciQuest: Milo's Mission
Introduction

Welcome to your quest to learn about science, or should I say, your SciQuest! This is a journey you are going to embark on with a bit of independence.

In *SciQuest 100: Milo's Mission*, you'll be learning about the basic building blocks of life, vertebrates, and our planet. Milo Pemberton will serve as your introduction for the week. You will learn about the topics in biology and earth science by reading, doing, and writing about science.

The weekly assignment sheets detail what you need to accomplish each week. These are broken down into the following categories. Here is a more detailed explanation of the three sections.

A Look At Each Unit

The Assignment Sheet

Easy unit will start with the unit's memory work. This is followed by a blank glossary for the vocabulary terms in the unit. After that, you will find the week's work. Here is a look at those sheets.

Read

In this section, you will find the chapter from Milo's story, *The PB Nanner Fluffer Test.* After that, you'll find pages from the following encyclopedias:

- *Usborne Science Encyclopedia*
- *Kingfisher Science Encyclopedia*
- *Usborne Illustrated Dictionary of Science (used in Units 1 and 2)*
- *Super Earth Encyclopedia (used in Unit 3)*
- *Everything You Need to Ace Science in One Big Fat Notebook*

Your teacher will tell you which ones to read and you can mark those on your assignment sheet.

Do

In this section, you will find your hands-on experiment for the week. There is also space for your teacher to assign an additional hands-on activity or for you to record one that you have chosen to do.

Write

In this section, you will find the vocabulary for the week and the sketch assignment. You will also have space to mark off the notes you have taken.

The last part of the assignment sheet is the key idea that you need to make sure you remember for the week.

The Experiment Sheets

The experiment sheets will help you record what you have done using the scientific method. Each sheet has the following:

Introduction

Each of your experiment sheets will begin with a question and an introduction. The introduction will give you some background knowledge for the experiment.

Hypothesis

In the hypothesis section, you need to predict the answer to the question posed in the lab.

Materials

In the materials section, you need to fill out what you used to complete the experiment.

Procedure

In the procedure section, you need to write a step-by-step account of what you did during your experiment. You need to provide enough detail so that someone else could read your report and replicate your experiment.

Observation and Results

In the observation section, you need to write what you saw and observed, as well as any results you measured.

Conclusion

Finally, in the conclusion section, you need to write whether your hypothesis was correct and any additional information you have learned from the experiment. If your hypothesis was not correct, discuss why with your teacher and then include why your experiment did not work on your experiment sheet.

The Sketch and Notes Sheets

These two sheets provide the space for you to record what you have learned for the week. They included a pre-drawn sketch and space to record what you have learned.

Sketch

Color the pre-drawn sketch and label it with the information given in the box on the sheet. Be sure to give the sketch a title as well.

Notes

On the notes sheets, you can write either a list of facts, an outline, and/or a narrative summary after you complete your reading assignments. Be sure to answer the questions provided on the sheet. If your assigned encyclopedias don't have the information, your teacher can provide you will the information.

How to Schedule this Study

SciQuest 100: Milo's Mission should take up to three hours per week to complete. You, along with your teacher, can choose whether to complete the work over five days or over two days. Below are two options for scheduling to give you an idea of how you can schedule your week:

Two-day-a-week Schedule

Day 1

Define the vocabulary, do the experiment, and complete the experiment sheet.

Day 2

Read the assigned pages and discuss them together, take your notes, and complete the sketch.

Five-day-a-week Schedule

Day 1

Do the experiment and complete the experiment sheet.

Day 2

Record the dates and define the vocabulary.

Day 3

Read the assigned pages and discuss them together and complete the sketch.

Day 4

Prepare the science report or an outline.

Day 5

Complete one of the other activities assigned by your teacher and work on the memory work.

List of Topics

Animal Basics Unit

- Plant Cells
- Animal Cells
- Microscopes
- Taxonomy
- Dichotomous Keys
- Migration
- Animal Navigation
- Adaptations
- Defenses
- Animal Diet (Herbivore, Carnivore, and Omnivore)

Vertebrates Unit

- Fish
- Amphibians
- Reptiles
- Birds
- Mammals
- Animal Reproduction
- Mitosis
- Meiosis

Scientist Biography Unit

- You will learn about a scientist of your choice.

Planet Earth Unit

- Inside the Earth
- Maps and Mapping
- Rivers
- Oceans
- Glaciers
- Natural Cycles
- Biomes and Habitats

SciQuest

Milo's Mission Lab Journal

Animal Basics Unit

What you'll study in this unit-

- O 1: The Cell
- O 2: Classification
- O 3: Migration
- O 4: Animal Diet
- O 5: Animal Defenses

Memory Work-

The Divisions of Life

Animal Basics Unit

Memory Work

Back when I was in school, I had a wonderful science teacher who insisted on finding ways for us to remember key facts. For the students in her class, including me, this wasn't optional. But for you it is. So, I'm going to share the memory work for this unit. You should read it, so you know what it is. But whether or not you have to memorize it, is up to your teacher!

Divisions of Life

Kingdom

Phylum

Class

Order

Family

Genus

Species

Note-

The following mnemonic can help you as you work on memorizing these:

King **P**hillip **C**an **O**nly **F**ind **H**is **G**reen **S**hoes.

Animal Basics
Vocabulary Sheet

Autotroph–________________________________

Camouflage – ________________________________

Carnivore–________________________________

Cell–________________________________

Classification – ________________________________

Hemisphere–________________________________

Herbivore–______________________________

Heterotroph – ______________________________

Kingdom–______________________________

Migration – ______________________________

Mitosis–______________________________

Omnivore–______________________________

Predator – ______________________________

Prey–____

Species–____

Lesson 1: The Cell

Assignment Sheet

Read

- O Read Chapter 1 in *The PB Nanner Fluffer Test*: "Milo Dresses up for Co-op."
- O Read the encyclopedia pages assigned by your teacher. Circle the ones your instructor wants you to do.
    - *Usborne Science Encyclopedia*: "Plant Cells" pp. 250-251, "Animal Cells" pp. 298-299
    - *Kingfisher Science Encyclopedia:* "Single-celled Organisms" p. 54
    - *Usborne Illustrated Dictionary of Science:* "The Structure of Living Things" pp. 238-241
    - *Everything You Need to Ace Science in One Big Fat Notebook:* "Cell Theory and Cell Structure" pp. 303-312
    - (Other) ____

Do

- O Do the "What do plant cells look like?" experiment. Fill out the experiment sheet as you do the experiment.
- O (Other) ____

Write

- O Look up cell and mitosis. Add the definitions to the unit vocabulary sheet.
- O Label the "Animal Cell" and "Plant Cell" sketches.
- O Write a list of facts or a paragraph about animal cells and plant cells on the pages with the sketches.
- O (Other) ____

____

> **Key Idea**
>
> **Cells are the basic building blocks of life.**

What do plant cells look like?

Introduction

We aren't going to dress up as plant or animal cells this week, like Milo did. But we are going to look at a few cells under the microscope. Remember that all living things are made up of tiny building blocks, called cells, which carry out the necessary functions of life for the animal and plant. The best way to see these cells is by using a microscope. In this experiment, you will view celery and onion cells.

Hypothesis

I think that the cells will look like ______________________________

____

Materials

You will need:

- Microscope
- Slides
- Onion skin
- Celery stalk
- Razor

Procedure

Note—If you have never used a microscope before, please ask the teacher to demonstrate how to use a microscope before beginning this experiment.

1. Make a wet mount slide of the onion skin and the celery stalk.

    0 Collect a thin slice of the sample and place it on the slide. (*Make sure the sample is very thin or else the cover slip will wobble and you won't get a very good view of the sample.*)

    0 Place one drop of water over the sample. (*Make sure not to use too much water or else the cover slip will float away and again you won't be able to see the sample.*)

    0 Place the cover slip at a 45 degree angle, with one edge touching the water, and let go.

2. Look at the slide under the microscope on low power (100x) and then on high power (400x). Draw what you see for each.

3. Next, look at the slide under the microscope on low power (100x) and then on high power (400x). Draw what you see for each.

Conclusion

Plant Cell Notes and Sketch

Your Turn: Write what you learned about plant cells.

Need help getting started?
Answer the following questions: What are three of the basic parts of a plant cell and what does each do? Are all plant cells the same? Why or why not? What are the two stages of cell division in a plant cell? What happens in each of them?

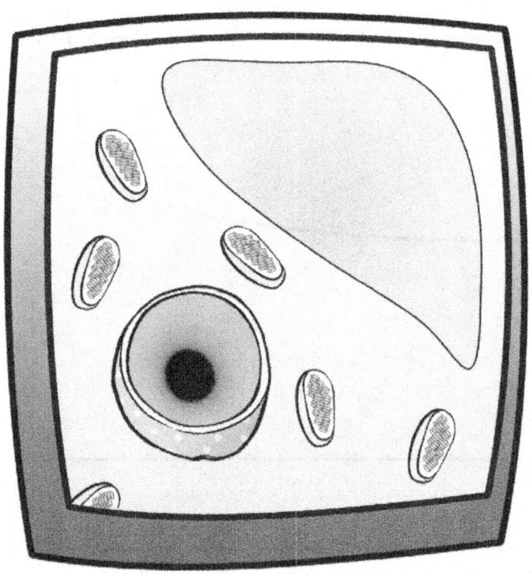

Label the following on the plant cell – cell wall, cell membrane, cytoplasm, vacuole, nucleus, chloroplasts

Bonus Question: How are plant cells and animal cells different?

Animal Cell Notes and Sketch

Your turn: Write what you learned about animal cells.

Need help getting started?
Answer the following questions: What are three of the basic parts of an animal cell and what does each do? Explain how cells come together to form organs.

Label the following on the animal cell – nucleus, nucleolus, mitochondria, vacuole, endoplasmic reticulum, ribosome, cell membrane, lysosome, cytoplasm

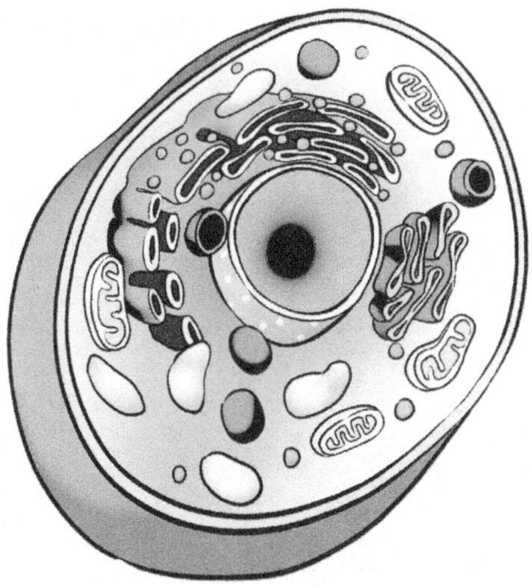

Lesson 2: Classification

Assignment Sheet

Read

- O Read Chapter 2 in *The PB Nanner Fluffer Test*: "Milo Unpacked... With Help."
- O Read the encyclopedia pages assigned by your teacher. Circle the ones your instructor wants you to do.
    - *Usborne Science Encyclopedia*: "Classifying Plants" pp. 294-295, "Classifying Animals" pp. 340-343
    - *Kingfisher Science Encyclopedia*: "Classification of Living Things" pp. 52-53
    - *Usborne Illustrated Dictionary of Science:* "The Classification of Living Things" pp. 340-341
    - *Everything You Need to Ace Science in One Big Fat Notebook:* "Organisms and Biological Classification" pp. 292-300
    - (Other) ____

Do

- O Do the "What kind of tree is it?" experiment. Fill out the experiment sheet as you do the experiment.
- O (Other) ____

Write

- O Look up classification, kingdom, and species. Add the definitions to the unit vocabulary sheet.
- O Label the "Divisions of Life" sketch.
- O Write a list of facts or a paragraph about migration on the pages with the sketches.
- O (Other) ____

  ____

Key Idea

All living things can be organized or grouped.

What kind of tree is it?

Introduction

You may not have a pile of boxes to sort as your stuff away like Milo did, but that doesn't mean you can't use a dichotomous key. This method can be used to identify a living thing, such as a tree. At each step, the user is given a question with two answers. The answers eventually lead to the identity of the living thing. In this experiment, you will be using a dichotomous key from the internet to identify an unknown leaf.

Hypothesis

I think the leaf is from the ____

____ tree.

Materials

You will need:

- Leaf from outside
- Dichotomous key for plants from the following website:
  https://treespnw.forestry.oregonstate.edu/

Procedure

1. Go outside and choose a leaf from a tree that you want to identify.
2. Once inside, write down what kind of tree you think the leaf is in the hypothesis section above.
3. Spend a few moments observing the leaf, making notes about its shape and size under the "Observations" section.
4. Use the dichotomous key from the website above to identify the leaf. (Although the website is for trees of the Pacific Northwest, it will help you identify trees from all over the USA and other countries. If you don't live in America and the key doesn't work for your leaf, just head over to your search engine of choice and type in "Tree Dichotomous Key for{the place you live}.)

Drawing or Picture of the Leaf

Results

____

____

____

____

Conclusion

____

____

____

____

Classification Sketch

Label the following in the boxes below – Kingdom, Phylum, Class, Order, Family, Genus, Species.
Add the title: Divisions of Life

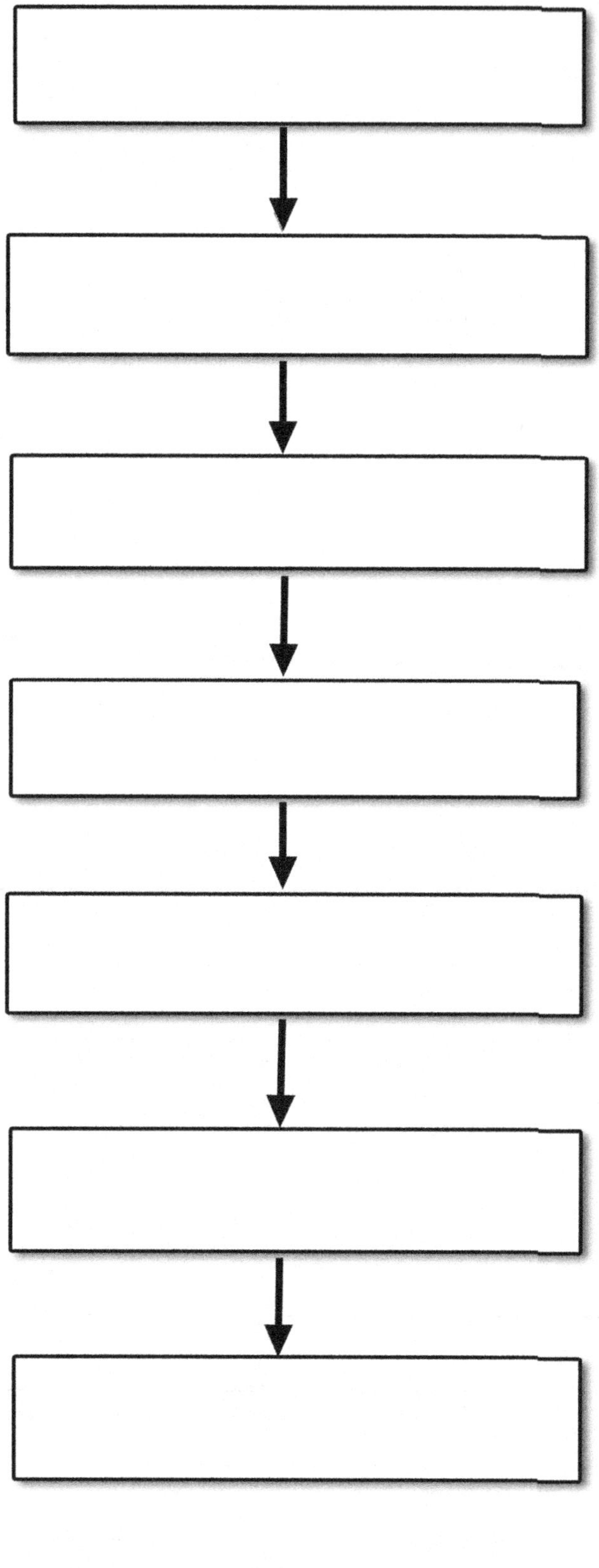

Classification Notes

Your turn: Write what you learned about classification.

Need help getting started?

Answer the following questions: What are the five main kingdoms? How do scientists classify living things? What are the two main divisions of the plant kingdom? What are the taxonomic ranks (or divisions of life)? What language is a biological name given in and how is it created?

Bonus Question: How has a deeper understanding of DNA changed Linnaeus's classification ideas?

Lesson 3: Migration

Assignment Sheet

Read

- O Read Chapter 3 in *The PB Nanner Fluffer Test*: "Milo Chuckled in Class."
- O Read the encyclopedia pages assigned by your teacher. Circle the ones your instructor wants you to do.
  - *Usborne Science Encyclopedia*: "Migration (section)" p. 329
  - *Kingfisher Science Encyclopedia*: "Migration" p. 93
  - *Usborne Illustrated Dictionary of Science:* "Section on Migration" p. 237
  - *Everything You Need to Ace Science in One Big Fat Notebook:* "Section on Migration" p. 370
  - (Other) ____________________

Do

- O Do the "Backyard Animal Observations" experiment. Fill out the experiment sheet as you do the experiment.
- O (Other) ____________________

Write

- O Look up hemisphere and migration. Add the definitions to the unit vocabulary sheet.
- O Label the "Migration Routes" sketch.
- O Write a list of facts or a paragraph about classification on the pages with the sketches.
- O (Other) ____________________

____

Key Idea

Animals migrate to breed and to find food.

Backyard Animal Observation

Introduction

Milo learned about migration in his co-op class. Often when we think of animals migrating, we think about those creatures that are far off in some distant space that we do not see. But there are animals the migrate through your area! In this experiment, you are going to observe your backyard for animals, then identify those animals, and finally do a bit of research to determine if they migrate.

Materials

You will need:

- A place to observe animals
- Access to the Internet or a local field guide

Procedure

1. Go outside.
2. Find a quiet place to sit and observe for about 30 minutes.
3. Record your observations on the following chart.
4. After the time is up, head inside to research the animals you observed using either the Internet or a local field guide. Begin by identifying the animals if you have not done so already. Then, mark down if these animals are seasonal to your area or if they live there year-round.

Observations

Observations	Identification	Migration

Observations	Identification	Migration

Conclusion

____

____

____

____

____

Migration Sketch

Label the following on the map below – On the map below trace the routes of the swallow, reindeer, monarch butterfly, arctic tern and humpback whale. Do each route in a different color or pattern and then add each route to the key in the box below the map. Add the title: Migration Routes.

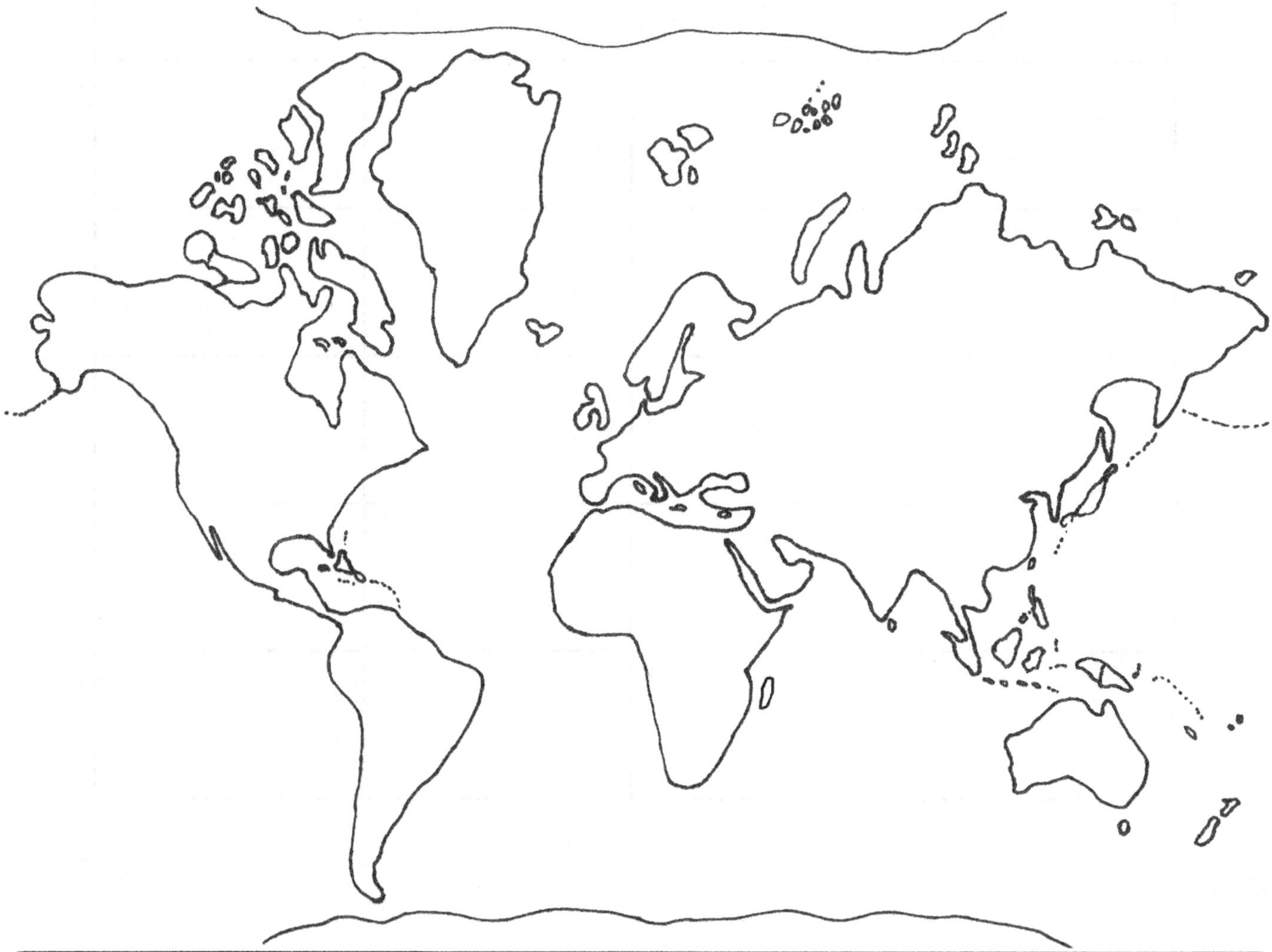

Map Key

Migration Notes

Your turn: Write what you learned about migration.

Need help getting started?
Answer the following questions: Why do animals migrate? How do animals find their way during migration? Which animals are the most common migrants? Name two ways that birds use to navigate.

Bonus Question: If you could migrate, where would you go? Why?

Lesson 4: Animal Diet

Assignment Sheet

Read

○ Read Chapter 4 in *The PB Nanner Fluffer Test*: "Milo Gave a Test."

○ Read the encyclopedia pages assigned by your teacher. Circle the ones your instructor wants you to do.

- *Usborne Science Encyclopedia*: "Feeding" pp. 310-311
- *Kingfisher Science Encyclopedia*: "Teeth and Digestion" pp. 312-313, "Food Chains" p. 435
- *Usborne Illustrated Dictionary of Science:* "Within An Ecosystem" pp. 234-235
- *Everything You Need to Ace Science in One Big Fat Notebook:* "Herbivore, Omnivore, and Carnivore" p. 362, "Feeding Relationship, Food Chains, and Food Webs" pp. 487-490
- (Other) ______________________________

Do

○ Do the "Owl Pellet" experiment. Fill out the experiment sheet as you do the experiment.

○ (Other) ______________________________

Write

○ Look up autotroph, heterotroph, carnivore, herbivore, and omnivore. Add the definitions to the unit vocabulary sheet.

○ Label the "Simple Food Chain" sketch.

○ Write a list of facts or a paragraph about animal defenses on the pages with the sketches.

○ (Other) ______________________________

____

Key Idea

Animal diet varies and each species is adapted for the food they eat.

Owl Pellet

Introduction

Milo and his friends learn about animal diet, but today we're going to see an animal's diet firsthand in an owl pellet. Unlike other birds, the food goes directly from the owl's mouth into its gizzard where digestion begins. The food is separated into digestible and indigestible parts. The indigestible parts, like the bones, fur, and sand, are compacted together into an oval-shaped pellet, which the owl regurgitates several hours after it eats. Since the owl swallows the prey virtually whole, most of the bones are intact and can be viewed quite easily in the pellet.

Materials

You will need:

- Gloves, Tweezers, Wooden stick (coffee stirrer or popsicle stick will work)
- Newspaper or paper towel
- Owl pellet

Procedure

1. Before you begin, lay out the newspaper or paper towels on your work surface and put on the gloves. Most pellets are sterilized, but you should still use these precautions anyways.
2. Open the owl pellet up and examine the outside. Look for fur, feathers, and other identifiable material.
3. Use the wooden stick to gently break the owl pellet up and observe what you see.
4. Use the tweezers to pick up some of the bones you can see.
5. Examine these bones up close to identify whether they came from a rodent, shrew, mole, and small bird. You may also see remnants of feathers, fur, and bits of sand and gravel. (See your instructor for a Bone Identification Chart.)
6. Write down the bones you find under the observations and results.
7. When you are done with your observations, take off the gloves, roll up your foil with the pellet and tools inside, and throw it all away.

Observations and Results

Conclusion

Animal Diet Sketch

Label the following on the sketch below – On the sketch of a simple food chain below, label the apex predator, herbivore prey, and food for hervivores and omnivores. Add the title: Simple Food Chain.

Animal Diet Notes

Your turn: Write what you learned about animal's diets.

Need help getting started?

Answer the following questions: What is phagocytosis? What are the different parts of an insect's mouth and what are they used for? How do carnivore's teeth differ from an herbivore's teeth? What can omnivores eat? What two adaptations do birds have in their digestive system to deal with solid food?

Bonus Question: Do you think it is easier to survive as an herbivore, a carnivore, or an omnivore?

Lesson 5: Animal Defense

Assignment Sheet

Read

- O Read Chapter 5 in *The PB Nanner Fluffer Test*: "Milo Accepted His Mission."
- O Read the encyclopedia pages assigned by your teacher. Circle the ones your instructor wants you to do.
    - *Usborne Science Encyclopedia*: "Sending Messages" pp. 318-319
    - *Kingfisher Science Encyclopedia*: "Adaptation and Defense" p. 95
    - *Everything You Need to Ace Science in One Big Fat Notebook:* "Relationships Among Populations" pp. 485-487
    - (Other) ____

Do

- O Do the "How does camouflage work" experiment. Fill out the experiment sheet as you do the experiment.
- O (Other) ____

Write

- O Look up hemisphere and migration. Add the definitions to the unit vocabulary sheet.
- O Label the "5 Types of Animal Defense" sketch.
- O Write a list of facts or a paragraph about classification on the pages with the sketches.
- O (Other) ____

____

Key Idea

Predators hunt for food. Prey (the food) evades by using defenses.

How does Camouflage Work?

Introduction

Milo's sketch perfectly depicted the predator/prey relationship. But animals don't like to become a predator's prey, so they have adapted defenses to protect themselves from harm. One of these defenses is to basically hide in plain sight, otherwise known as camouflage. In today's experiment, you are going to use bits of paper to see how this animal defense works.

Hypothesis

I think that color and shape (do / do not) play a role in camouflage.

Materials

You will need:

- Newspaper
- Plain paper
- Black and green construction paper

Procedure

1. Cut out several of the same shapes from the newspaper, plain paper, black construction paper, and green construction paper.
2. Lay out a sheet of newspaper on the table. Then lay the shapes on top of the paper. Observe which shapes are easier to find. (*You can even have a partner come in and try to identify the number and types of shapes that are on the newspaper.*)
3. After that, use what you observed and learned in step 2 to cut several different shapes from the paper that was the hardest to see. Again, lay out these shapes to test which is the hardest to find. (*Once more, you can have a partner come in and try to identify the number of shapes that are on the newspaper.*)
4. Finally, move the shapes around. Ask yourself, is there a better position or place that will camouflage the shapes more adequately?

Observations and Results

Conclusion

Migration Sketch

Label the animals below with the type of defense it uses. You should use: mechanical defense, chemical defense, behavioral defense, electrical defense, and visual defense. Add the title: 5 Types of Animal Defenses

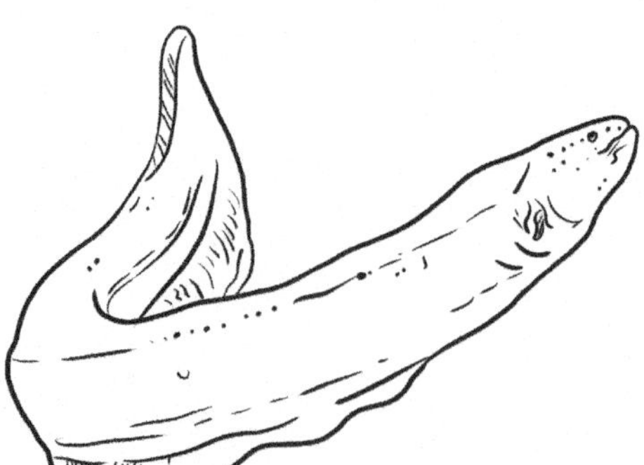

MIGRATION NOTES

YOUR TURN: Write what you learned about animal defense and adaptation.

NEED HELP GETTING STARTED?
Answer the following questions: What is the difference between adaptation and defense? What are the five main types of defensive weapons (give examples of each)? What are the two different types of adaptations? What are genotypic adaptations? Phenotypic adaptations?

SciQuest

Milo's Mission Lab Journal

Vertebrates Unit

What you'll study in this unit-

- 6: Fish
- 7: Birds
- 8: Amphibians
- 9: Reptiles
- 10: Mammals

Memory Work-

Basic Classes of Vertebrates

Vertebrates Unit

Memory Work

For this unit, we'll be memorizing a few basics facts about the animals with backbones.

Basic Classes of Vertebrates

Fish
Cold-blooded animals that live in water; they are covered with scales and breathe through gills.

Amphibians
Cold-blooded animals that live on land and in the water; they have soft skin.

Reptiles
Cold-blooded animals that lay eggs; they are covered with scales.

Birds
Warm-blooded animals that lay eggs; they have feathers and wings.

Mammals
Warm-blooded animals that feed their young milk; they are covered with fur.

Vertebrates

Vocabulary Sheet

Amphibian – ______________________________

____

____

Clutch – ______________________________

____

____

Crop – ______________________________

____

____

Ectotherm – ______________________________

____

____

Endotherm – ______________________________

____

____

Fledgling – ______________________________

____

____

Gills – ____

____

____

Gizzard – ____

____

____

Mammals – ____

____

____

Mammary gland – ____

____

____

Monotreme– ____

____

____

Reptile – ____

____

____

Scales – ____

____

____

Swim bladder – ____

____

____

Tadpole – ____

____

____

Vertebrate – ____

____

____

Lesson 6: Fish

Assignment Sheet

Read

- ○ Read Chapter 6 in *The PB Nanner Fluffer Test*: "Milo Went on a Field Trip."
- ○ Read the encyclopedia pages assigned by your teacher. Circle the ones your instructor wants you to do.
    - *Usborne Science Encyclopedia*: "Movement in Water" pp. 304-305
    - *Kingfisher Science Encyclopedia*: "Fish" pp. 78-79
    - *Usborne Illustrated Dictionary of Science:* "Swimmers" p. 269, "Gills" p. 272
    - *Everything You Need to Ace Science in One Big Fat Notebook:* "Fish" pp. 356-357
    - (Other) ____________________

Do

- ○ Do the "Which fish will float higher?" experiment. Fill out the experiment sheet as you do the experiment.
- ○ (Other) ____________________

Write

- ○ Look up gills, swim bladder, and vertebrate. Add the definitions to the unit vocabulary sheet.
- ○ Label the "Anatomy of a Fish" sketches.
- ○ Write a list of facts or a paragraph about fish on the pages with the sketches.
- ○ (Other) ____________________

____

Key Idea

Fish are cold-blooded animals that live in the water.

Which fish will float higher?

Introduction

Milo and his friends played a game of "Which Fish is it?" at the aquarium. Fish are a part of the phyla chordata, which means that they have a backbone. They also have a unique organ called a swim bladder. The swim bladder is an organ that helps to control the fish's buoyancy. Buoyancy is the ability to float in a liquid. In this experiment, we are going to look at how the amount of air in the swim bladder (or balloon in our case) affects the buoyancy.

Hypothesis

I believe that the balloon with the ____

amount of air in it will float the highest.

Materials

You will need:

- 1 Large clear glass jar or bowl
- 3 Small balloons
- 3 Small marbles
- Ruler
- Water

Procedure

1. Fill the glass cup or bowl with cool water.
2. Label each balloon with #1, #2, or #3 and stretch each balloon so that you can place a marble in it.
3. Tie balloon #1 off with just the marble in it and no air. Blow up balloon #2 a little bit and tie it off. Blow up balloon #3 about halfway and tie it off.
4. Place the balloons gently in the water and let them come to rest (about one minute).
5. Then measure where each one stays and record the measurements. (*Be sure to include the units.*)

Observations and Results

Balloon #1 floated ____________________ high

Balloon #2 floated ____________________ high

Balloon #3 floated ____________________ high

____

____

____

____

____

Conclusion

____

____

____

____

____

Fish Sketch

Label the following on the sketch below – mouth, eye, brain, spinal cord, lateral line, swim bladder, heart, gills, dorsal fin, caudal fin, anal fin, pelvic fin, pectoral fin. Add the title: Anatomy of Fish

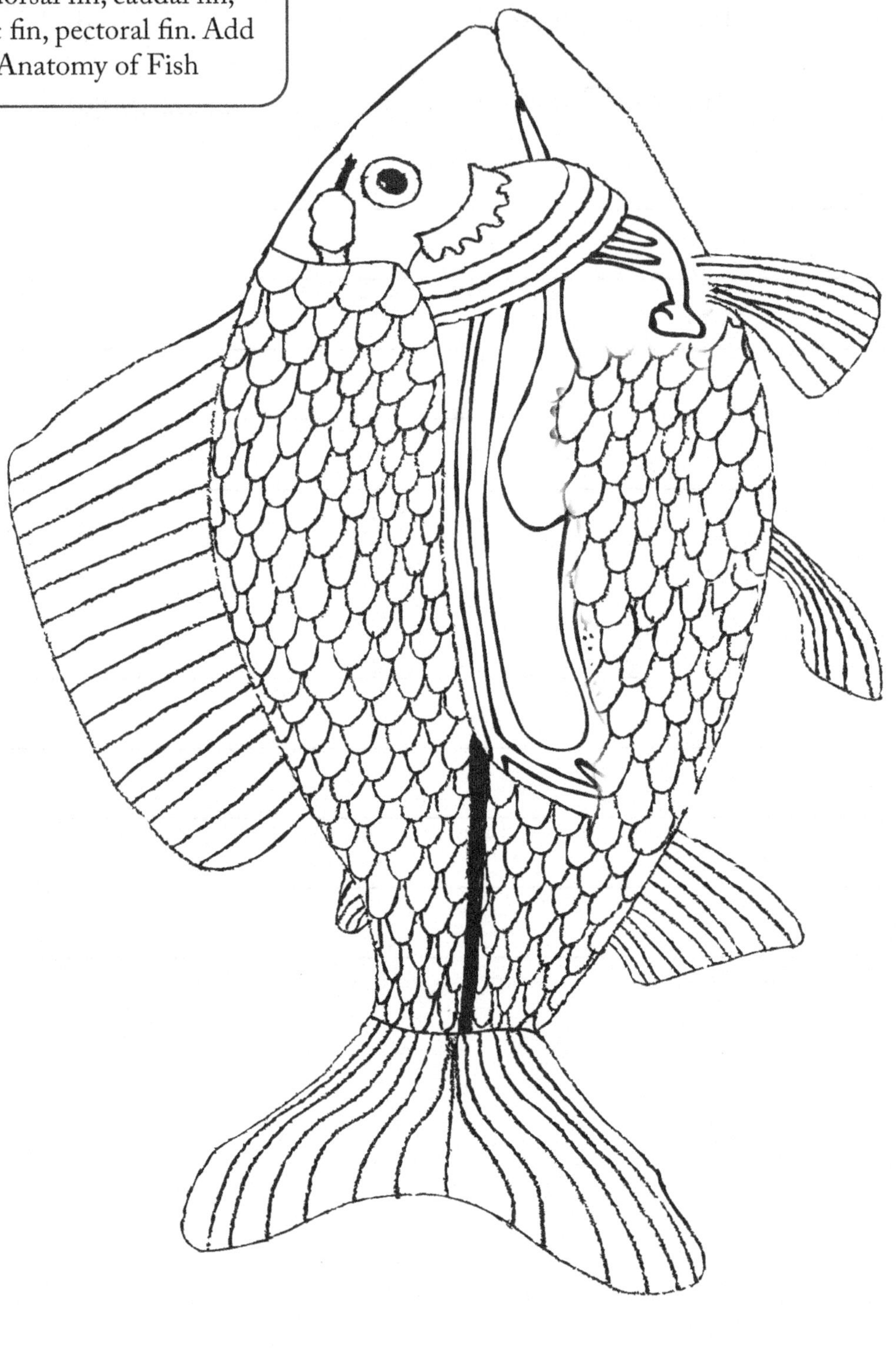

Fish Notes

Your turn: Write what you learned about fish.

Need help getting started?
Answer the following questions: What are the three types of fish and their main characteristics? How do saltwater and freshwater fish get rid of excess water from their bodies? How do fish usually reproduce? What is the lateral line and how does a fish use it? Explain how a fish breathes in the water.

Bonus Question: Why do you think fish lay so many eggs?

Lesson 7: Birds

Assignment Sheet

Read

- O Read Chapter 7 in *The PB Nanner Fluffer Test*: "Milo Ate Watermelon."

- O Read the encyclopedia pages assigned by your teacher. Circle the ones your instructor wants you to do.
  - *Usborne Science Encyclopedia*: "Movement in Air" pp. 306-307
  - *Kingfisher Science Encyclopedia*: "Birds" pp. 84-85
  - *Usborne Illustrated Dictionary of Science:* "Feathers" p. 267, "Flyers" p. 269
  - *Everything You Need to Ace Science in One Big Fat Notebook: :* "Birds" p. 360
  - (Other) ____

Do

- O Do the "Which beak?" experiment. Fill out the experiment sheet as you do the experiment.

- O (Other) ____

Write

- O Look up clutch, crop, gizzard, and fledgling. Add the definitions to the unit vocabulary sheet.

- O Label the "Three Types of Bird Feet" sketch.

- O Write a list of facts or a paragraph about birds on the pages with the sketches.

- O (Other) ____

____

Key Idea

Birds are warm-blooded animals with feathers and wings.

Which Beak?

Introduction

Milo and his friends did an experiment about bird beaks in their co-op class. And now it's your turn. Remember that all birds have bills or beaks that vary in shape and size. Birds use their beaks to obtain and manipulate their food. They also use their beaks for preening their feathers, for building their nests, and even for climbing! In this experiment, you are going to test to see how the size and shape of a beak relates to the bird's diet.

Hypothesis

Match the beak to the food you think it will best be able to pick up:

Long, large beak (chopsticks)	sugar water or honey	
Short, light beak (tweezers)	unshelled peanuts	
Short, strong beak (pliers)	bread	seeds
Long, slender beak (eyedropper)	worms	raisins

Materials

You will need:

- Chopsticks, Tweezers, Pliers, Eye dropper
- Sugar, Water, Gummy worms, Peanuts, Seeds, Raisins
- Plates

Procedure

1. Prepare five plates: one for each of the different types of food.
2. Determine the scale you would like to use when rating ease of pickup. (*For example, the scale range could be from 1 to 10, with "1" standing for the easiest and "10" standing for the hardest.*)
3. Fill in the scale range on the chart in the observations section.
4. Use the chopsticks to try to pick up each of the different types of food. Rate the ease of pickup on the chart on the experiment sheet.
5. Repeat using the tweezers, pliers, and eye dropper.

Observations

Scale Range: ______________________________________

Easiest = ____________________ Hardest = ____________________

	Bread	Gummy Worms	Unshelled Peanuts	Sugar Water	Seeds	Raisins
Chopsticks						
Tweezers						
Pliers						
Eye Dropper						

Results

____

____

____

Conclusion

____

____

____

____

____

Birds Sketch

Label the following on the sketch below – hooked claws and talons from a bird of prey, a backward-pointing toe from a perching bird, webbed feet from a swimming bird. Add the title: Three Types of Bird Feet.

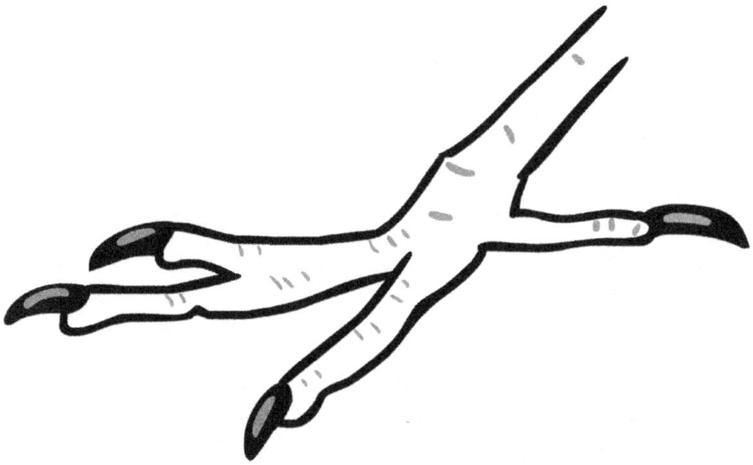

Birds Notes

Your turn: Write what you learned about birds.

Need help getting started?
Answer the following questions: What are three characteristics that enable birds to be so efficient at flight? What do birds use their feathers for? Why do birds regularly preen? What is the syrinx and what is its purpose? How do all birds reproduce? Name two adaptations that birds have for flight.

Bonus Question: If you had to make a bird's nest to protect your eggs, what materials would you use and where would you put the nest?

Lesson 8: Amphibians

Assignment Sheet

Read

- O Read Chapter 8 in *The PB Nanner Fluffer Test*: "Milo Walked in the River."
- O Read the encyclopedia pages assigned by your teacher. Circle the ones your instructor wants you to do.
    - *Usborne Science Encyclopedia*: "Amphibians" pp. 80-81
    - *Kingfisher Science Encyclopedia*: "Life Cycles" pp. 328-329
    - *Everything You Need to Ace Science in One Big Fat Notebook:* "Amphibians" p. 358
    - (Other) ____

Do

- O Do the "Frog Dissection" experiment.
- O (Other) ____

Write

- O Look up amphibian and tadpole. Add the definitions to the unit vocabulary sheet.
- O Label the "Life Cycle of a Frog" sketch.
- O Write a list of facts or a paragraph about amphibians on the pages with the sketches.
- O (Other) ____

____

Key Idea

Amphibians are cold-blooded animals that live on land and in water.

AMPHIBIAN SKETCH

Label the following on the sketch below – embryos develop in eggs laid by mature adults, tadpole hatches from the egg, tadpoles grows, tadpole develops legs, young moves from the water onto the land, adult frog matures. Add the title: Life Cycle of a Frog

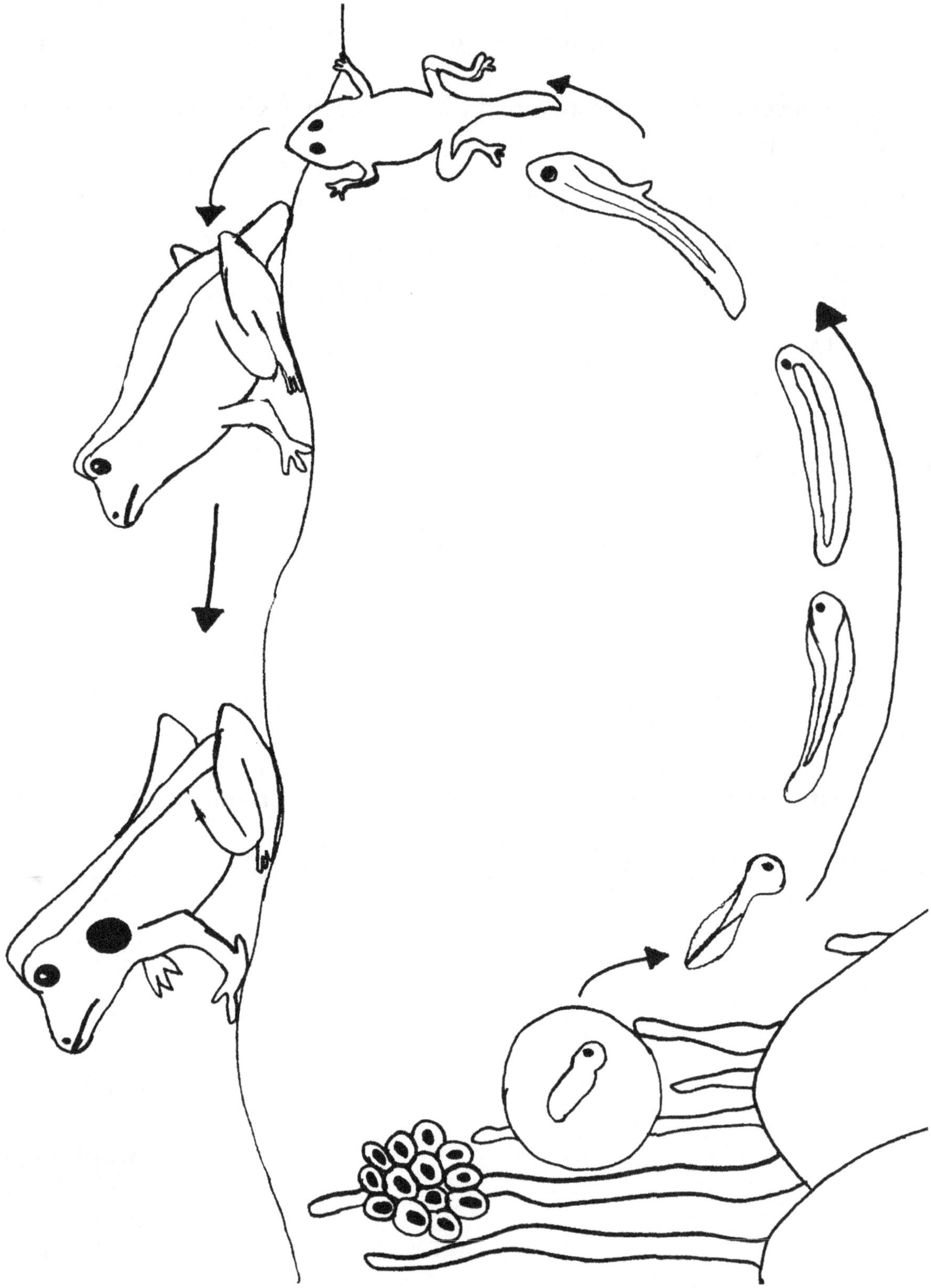

Amphibian Notes

Your turn: Write what you learned about amphibians.

Need help getting started?
Answer the following questions: What are three characteristics of amphibians? What are the three main groups of amphibians? How can you tell the difference between frogs and toads? Briefly describe the typical amphibian life cycle.

Bonus Question: Why might amphibians be known as "indicator species" for the health of an ecosystem?

Lesson 9: Reptiles

Assignment Sheet

Read

- O Read Chapter 9 in *The PB Nanner Fluffer Test*: "Milo Faced His PE Nemesis."
- O Read the encyclopedia pages assigned by your teacher. Circle the ones your instructor wants you to do.
  - *Usborne Science Encyclopedia*: "Body Covering" pp. 302-303
  - *Kingfisher Science Encyclopedia*: "Reptiles" pp. 82-83
  - *Everything You Need to Ace Science in One Big Fat Notebook:* "Reptiles" p. 359
  - (Other) ____

Do

- O Do the "Do I use only my tongue to taste things?" experiment. Fill out the experiment sheet as you do the experiment.
- O (Other) ____

Write

- O Look up ectotherm, reptile, and scales. Add the definitions to the unit vocabulary sheet.
- O Label the "Chameleon Defense" sketch.
- O Write a list of facts or a paragraph about reptiles on the pages with the sketches.
- O (Other) ________________________________

  ____

Key Idea

Reptiles are cold-blooded animals covered with scales.

Do I use only my tongue to taste things?

Introduction

Milo learned about migration in his co-op class. Often when we think of animals migrating, we think about those creatures that are far off in some distance space that we do not see. But there are animals the migrate through your area! In this experiment, you are going to observe your backyard for animals, then identify those animals, and finally do a bit of research to determine if they migrate.

Hypothesis

Do I use only my tongue to taste things? Yes No

Materials

You will need:

- Clothespin
- Blindfold
- 5 Plates
- 5 Pieces of bread with different edible spreads on them (such as garlic, cinnamon/ sugar, butter, peanut butter, and plain)

Procedure

1. Prepare the bread with the various edible spreads on five different plates and number the samples from #1-5.
2. Use the clothespin to close your nose. Then have your partner blindfold you and mix up the plates.
3. Have your partner feed each food to you, while you try to guess what each one is.
4. Remove the clothespin. Then taste each of the samples again.
5. Record the results.

Observations and Results

	Sample	My guess with the clothespin & blindfold on	My guess with the clothespin & blindfold off
1			
2			
3			
4			
5			

Conclusion

____

____

____

____

____

Reptile Sketch

Label the following on the sketch below – A chameleon's defense is to change its color. Color the first chameleon pale green with yellow stripes; label it "normal chameleon." Color the second chameleon dark brown with light brown stripes; label it "chameleon when threatened." Add the title: A Chameleon's Defense.

Reptiles Notes

Your turn: Write what you learned about reptiles.

Need help getting started?
Answer the following questions: What are some characteristics of a reptile? What are the four main groups of reptiles? How do reptiles control their body temperature? What are some characteristics of crocodilians? What characteristic is unique to tortoises and turtles? What are some characteristics of snakes? What do some snakes and lizards do that most reptiles do not?

Bonus Question: How do you think being cold-blooded changes what an animal does during the day?

Lesson 10: Mammals

Assignment Sheet

Read

- ○ Read Chapter 10 in *The PB Nanner Fluffer Test*: "Milo Navigated a Corn Maze."
- ○ Read the encyclopedia pages assigned by your teacher. Circle the ones your instructor wants you to do.
    - *Usborne Science Encyclopedia*: "Movement on Land" pp. 308-309
    - *Kingfisher Science Encyclopedia*: "Mammals" pp. 86-87, "Animal Reproduction" pp. 88-89
    - *Usborne Illustrated Dictionary of Science:* "Mitosis (cell division)" pp. 240-241, "Animal Reproduction and Development" pp. 276-277
    - *Everything You Need to Ace Science in One Big Fat Notebook:* "Cell Division and Mitosis" pp. 321-323, "Mammals" pp. 360-361
    - (Other) ____________________

Do

- ○ Do the "Which type of fur keeps you warmer?" experiment. Fill out the experiment sheet as you do the experiment.
- ○ (Other) ____________________

Write

- ○ Look up endotherm, mammals, mammary gland, and monotreme. Add the definitions to the unit vocabulary sheet.
- ○ Label the "Mitosis" sketch.
- ○ Write a list of facts or a paragraph about mammals on the pages with the sketches.
- ○ (Other) ____________________

____

Key Idea

Mammals are warm-blooded animals with fur.

Which type of fur keeps you warmer?

Introduction

Milo and his friends stroked the soft fur of a bunny at the fall festival, but fur is more than something soft to pet. Remember that all mammals are warm-blooded, which means that they are able to maintain a constant internal body temperature no matter what the outside temperature is. Mammals use their food to make energy that helps keep them warm. They also have fur to serve as a barrier to cold air and water. In this experiment, we will look at several types of fur to see which does the best job of keeping the cold out.

Hypothesis

I think that ______________________ will keep a mammal the warmest.

Thin fur (felt only) Thick fur (cotton balls)

Very Thick fur (felt & cotton balls) No fur

Materials

You will need:

- Felt, Cotton balls, Water, 1 Large cup, 4 Small cups, Instant read thermometer, Foil

Procedure

1. Line the outside of cup #1 with felt and fill three-quarters of the way with water and label it.
2. Next, take the large cup and place a few cotton balls at the bottom of the cup.
3. Then place cup #2 on top of the cotton balls, and fill in the space with more cotton balls. Then fill cup #2 three-quarters of the way with water and label it.
4. Glue a layer of cotton ball onto a layer of felt and wrap it around cup #3, and then cover it with another layer of felt. Fill cup #3 three-quarters of the way with water and label it.

5. Fill cup #4 three-quarters of the way with water and label it. This is the control.
6. Take the initial temperature of all four cups, record, and cover with foil.
7. Place all four cups into the freezer.
8. Measure and record the temperatures every thirty minutes over a period of three hours.
9. Draw conclusions.

Observations and Results

Temperature	Initial	After 30 min	After 60 min	After 90 min	After 120 min	After 150 min	After 180 min
Cup #1							
Cup #2							
Cup #3							
Cup #4							

Conclusion

Mammals Sketch

Label the sketch below with the following informaiton: chromosomes change into thick coils, chromosomes line up in the center, the chromosomes pairs split apart, cells pinch apart, daughter cells grows. Add the title: Mitosis

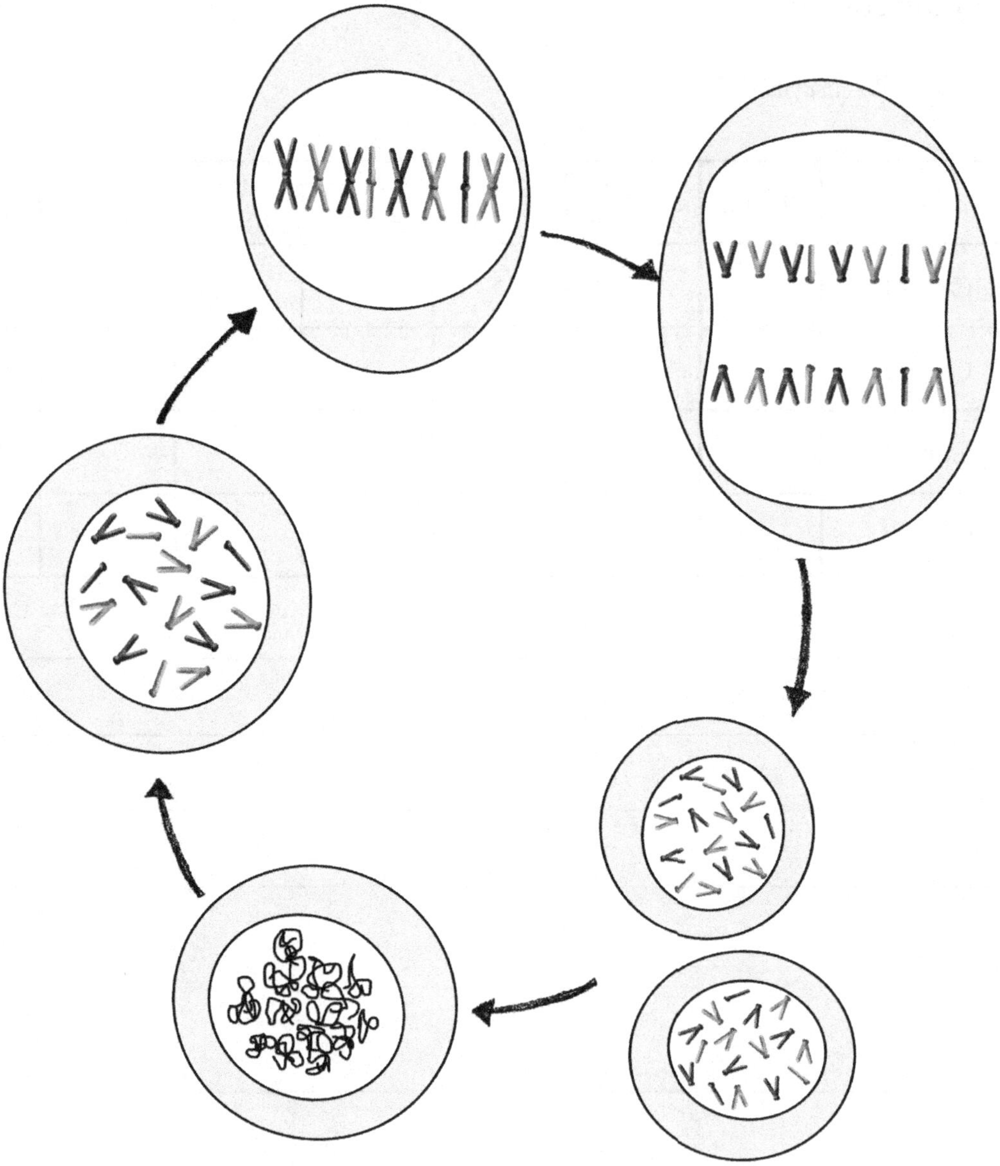

Mammals Notes

Your turn: Write what you learned about mammals.

Need help getting started?
Answer the following questions: What are two other characteristics of most mammals? What are the three groups of mammals? Explain how the three groups of mammals differ in the way that they reproduce. What two factors have contributed to the success of mammals? How is animal's lifespan related to growth and reproduction? Which group of chordates takes the longest to reproduce? Why? What are clones?

SciQuest

Milo's Mission Lab Journal

Scientist Biography Unit

What you'll study in this unit-

- O 11: The Scientist of Your Choice

Memory Work-

None

Lesson 11: Scientist Biography

Assignment Sheet

Read

- O Read Chapter 11 of *The PB Nanner Fluffer Test*: "Milo Visited the Library."
- O Choose and read a scientist biography.
  - (Book Title) ____

Do

- O (Other) ____

Write

- O Create a poster or write a report about the scientist you choose.
- O (Other) ______________________________

  ____

Scientist Biography

Step 1: Choose A Scientist

During the next few weeks, you are going to be researching and learning more about a scientist that has contributed to science. You need to begin your scientist biography project by choosing which scientist you will research. You can choose someone from the field of biology, like Galen, Robert Hooke, Carl Linnaeus, Gregor Mendel, Alexander Fleming, or Jacques Cousteau, or someone that has interested you.

The scientist I am going to study is:

____

Step 2: Research the Scientist

Once you have chosen the scientist you would like to study, you can begin your research. Begin by looking for a biography on your chosen scientist at the library. Then, look for articles on the chemist in magazines, newspapers, encyclopedias, or on the Internet. You will need to know the following about your scientist to write your report:

- 0 Biographical information on the scientist (where he or she was born, his or her parents, siblings, and how he or she grew up)
- 0 The scientist's education (where he or she went to school, what kind of student he or she was, what he or she studied, and so on)
- 0 His or her scientific contributions (the research that he or she participated in, any significant discoveries he or she made, and the state of the world at the time of his or her contributions)

As you read over the material you have gathered, be sure to write down any facts you glean in your own words. You can do this on a separate sheet or on separate index cards - just make sure you have at least five pieces of information for each category. And don't forget to record a list of the resources you have used for your research.

Step 3: Create an Outline

Now that your research is completed, you are ready to begin the process of writing a report on your chosen scientist. You are going to organize the notes you took during step two into a formal outline which you will use next week to write the rough draft of your report. Use the outline template provided below as a guide. You should include information on why you chose the particular scientist in your introduction section. For the conclusion section of the outline, you need to include why you believe someone else should learn about your chosen scientist and your impression of the scientist (i.e. *Did you like the scientist? Do you feel that they made a significant impact on their field of science?*).

I. Introduction and Biological Information on the Scientist

A.

B.

C.

D.

E.

F.

II. The Scientist's Education

A.

B.

C.

D.

E.

F.

III. The Scientist's Contributions

A.

1.

2.

3.

4.

B.

1.

2.

3.

4.

C.

1.

2.

3.

4.

IV. Conclusion

A.

B.

C.

D.

E.

Step 4: Write a Rough Draft

Finally you are ready to take your outline and turn it into a rough draft. Simply take the points on your outline and combine or add in sentence openers and transitions to create a cohesive paragraph. Here's what your rough draft should look like:

0 Paragraph 1 (from outline point I): introduce the scientist;

0 Paragraph 2 (from outline point II): tell about the scientist's education;

- 0 Paragraph 3-5 (from outline point III): share the scientist's contributions (one paragraph for each contribution);
- 0 Paragraph 6 (from outline point IV): share your thoughts on the scientist and why someone should learn about him or her.

You can choose to hand write or type up your rough draft on a separate sheet of paper. However, keep in mind that you will need a typed version for step five.

Step 5: Revise to Create a Final Draft

Now that you have a typed, double-spaced rough draft, look over it one more time to make any changes you would like. Then, have your teacher or one of your peers look over the paper for you to correct any errors and bring clarity to any of the unclear sections. Once this is complete, make the necessary changes to your paper to create your final draft. Print out your final paper to share or use it to create a book or poster on the scientist.

SciQuest

Milo's Mission Lab Journal

Planet Earth Unit

What you'll study in this unit-

- O 12: Inside the Earth
- O 13: Maps and Mapping
- O 14: Rivers
- O 15: Oceans
- O 16: Glaciers
- O 17: Biomes and Habitats
- O 18: Natural Cycles

Memory Work-
The World's Major Seas and Oceans

Planet Earth Unit

Memory Work

For this unit, we'll be memorizing the major lines of longitude and latitude, along with the major seas and oceans around the world.

Major Lines on the Globe

1. Prime Meridian 2. Equator 3. Tropic of Cancer 4. Tropic of Capricorn

The World's Major Seas and Oceans

1. Arctic Ocean
2. Pacific Ocean
3. Atlantic Ocean
4. Indian Ocean
5. Southern Ocean
6. Mediterranean Sea
7. Red Sea
8. Black Sea
9. Caribbean Sea
10. Gulf of Mexico
11. Hudson Bay
12. Bering Sea
13. Tasman Sea
14. Coral Sea
15. Bay of Bengal
16. Arabian Sea
17. North Sea

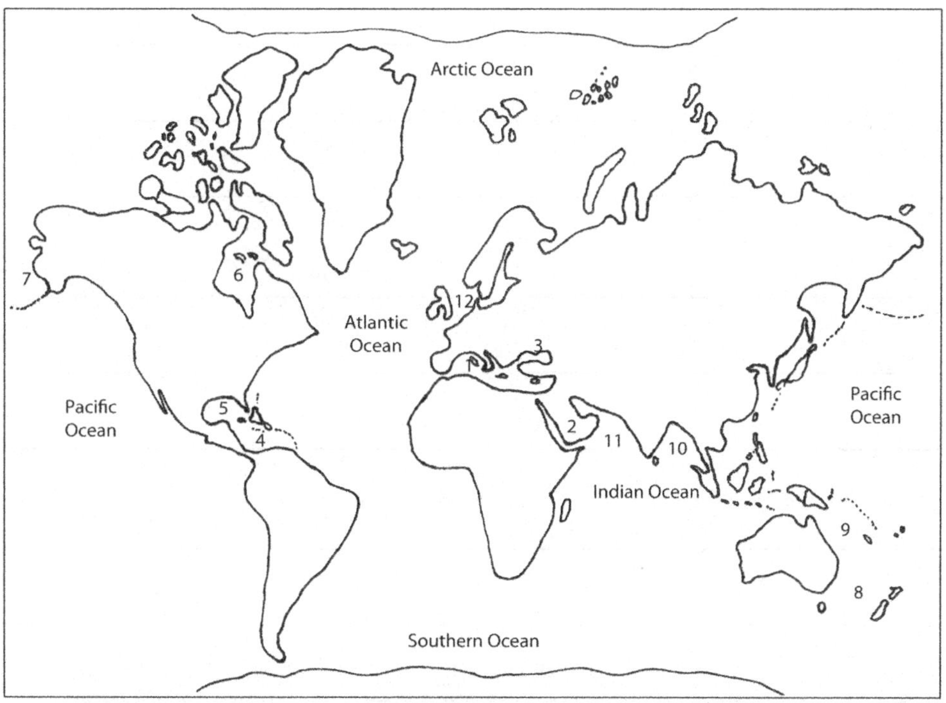

1. Mediterranean Sea
2. Red Sea
3. Black Sea
4. Caribbean Sea
5. Gulf of Mexico
6. Hudson Bay
7. Bering Sea
8. Tasman Sea
9. Coral Sea
10. Bay of Bengal
11. Arabian Sea
12. North Sea

Planet Earth
Vocabulary Sheet

Biome – ______________________________

____

____

Cartographer – ______________________________

____

____

Coast – ______________________________

____

____

Delta – ______________________________

____

____

Deposition – ______________________________

____

____

Erratics – ______________________________

____

____

Estuary – ____

____

____

Glacier – ____

____

____

Greenhouse gas – ____

____

____

Lines of Latitude (parallels) – ____

____

____

Lines of Longitude (meridians) – ____

____

____

Mantle – ____

____

____

Moraine – ____

____

____

Natural cycle – ____

____

____

Oceanic ridge – ____

____

____

Oceanic trench – ____

____

____

Source – ____

____

____

Tides – ____

____

____

Lesson 12: Inside the Earth

Assignment Sheet

Read

- ○ Read Chapter 12 in *The PB Nanner Fluffer Test*: "Milo Shared a Cartoon."
- ○ Read the encyclopedia pages assigned by your teacher. Circle the ones your instructor wants you to do.
    - *Usborne Science Encyclopedia*: Earth's Structurere" pp. 180-181
    - *Kingfisher Science Encyclopedia:* "Earth's Structure" pp. 8-9
    - *Super Earth Encyclopedia:* "Layered Planet" pp. 14-15
    - *Everything You Need to Ace Science in One Big Fat Notebook:* "The Earth's Structure and Composition" pp. 235-236
    - (Other) ____

Do

- ○ Do the "Does the molten layer in Earth's core affect its movement?" experiment. Fill out the experiment sheet as you do the experiment.
- ○ (Other) ____

Write

- ○ Look up mantle. Add the definitions to the unit vocabulary sheet.
- ○ Label the "Inside the Earth" sketches.
- ○ Write a list of facts or a paragraph about what's inside the Earth on the pages with the sketches.
- ○ (Other) ____

____

Key Idea

The Earth consists of three main layers: the crust, the mantle, and the core.

Does the molten layer in Earth's core affect its movement?

Introduction

Milo and his friends made a model of the Earth. The Earth consists of three main layers: the crust, the mantle, and the core. Some of these layers are solid, some of them are molten liquid. Each layer varies in thickness and consists of different elements. In this experiment, you are going to see how the composition of our Earth affect its motion.

Hypothesis

I think the (raw / hard-boiled) egg will spin for longer.

Materials

You will need:

- Permanent marker
- 1 Raw egg
- 1 Hard-boiled egg

Procedure

1. Mark the raw egg with a #1 and the hard-boiled egg with a #2.
2. Set the raw and the hard-boiled egg on the counter for twenty minutes to allow them to come up to room temperature.
3. Place both the eggs on the counter and try to spin each egg.
4. Observe how each one moves and record the time each egg spins on the chart.

Observations

	Raw Egg	Hard-boiled Egg
Spin Time		

Results

Conclusion

Inside the Earth Sketch

Label the following on the sketch below – inner core, outer core, mantle, crust, ocean, and continent. Add the title: Inside the Earth

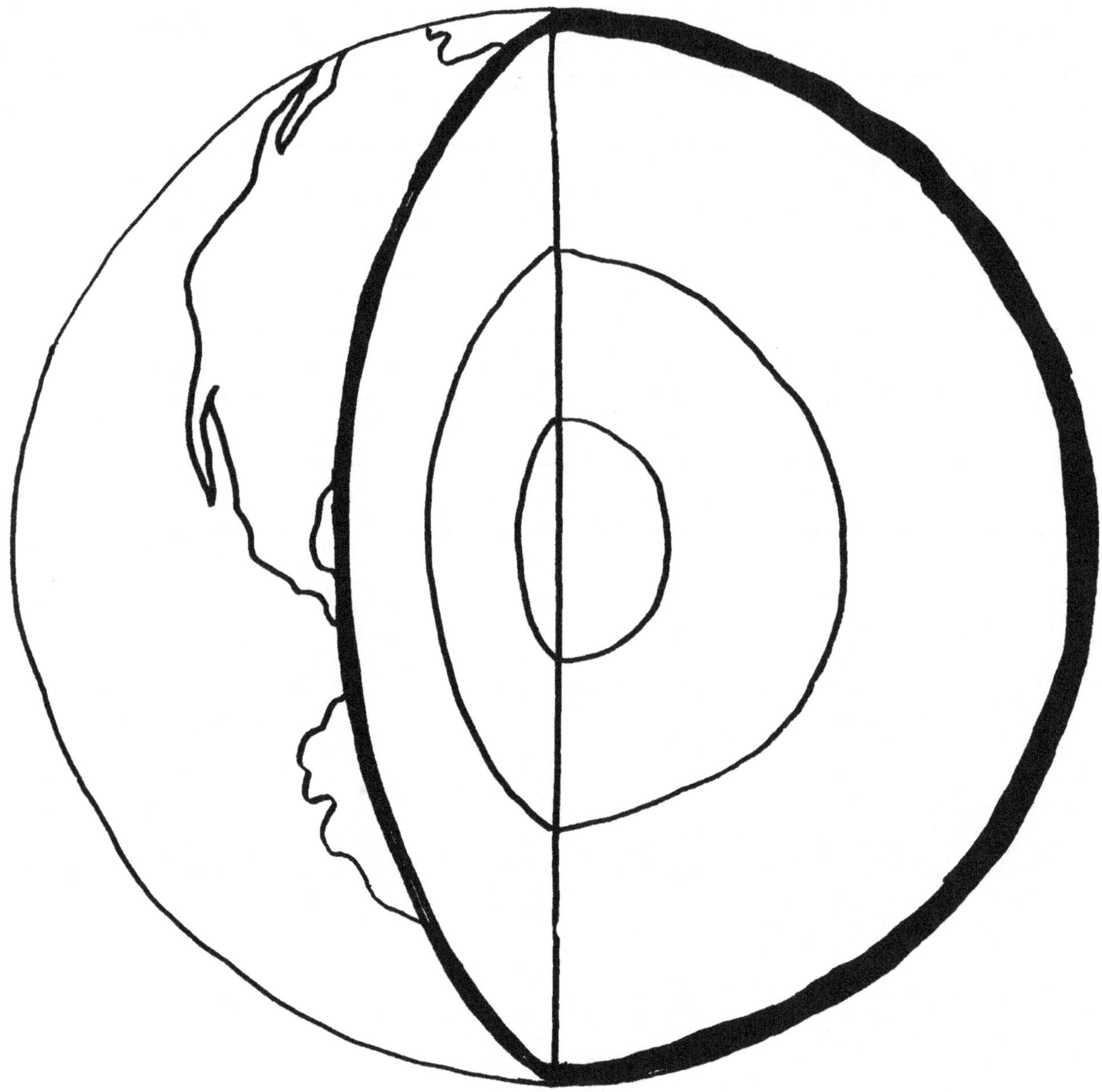

Inside the Earth Notes

Your turn: Write what you learned about what's inside the Earth.

Need help getting started?
Answer the following questions: What are the three main layers of the Earth? How is the Earth's magnetic field created and why is it important? What causes seismic activity on the Earth? What have seismic scans of the Earth revealed about the mantle?

Bonus Question: How would things change on Earth if the inside was all solid rock?

Lesson 13: Maps

Assignment Sheet

Read

- O Read Chapter 13 in *The PB Nanner Fluffer Test*: "Milo Solved the Riddle."
- O Read the encyclopedia pages assigned by your teacher. Circle the ones your instructor wants you to do.
    - *Kingfisher Science Encyclopedia*: "Maps and Mapping" pp. 46-47
    - *Everything You Need to Ace Science in One Big Fat Notebook:* "Topographic Maps" p. 256
    - (Other) ______________________________

Do

- O Do the "Flat vs. Sphere" experiment. Fill out the experiment sheet as you do the experiment.
- O (Other) ______________________________

Write

- O Look up cartographer, lines of longitude, and lines of latitude. Add the definitions to the unit vocabulary sheet.
- O Label the "Lines on the Globe" sketch.
- O Write a list of facts or a paragraph about maps and mapping on the pages with the sketches.
- O (Other) ______________________________

  ____

Key Idea

Maps show physical information about an area.

Flat vs. Sphere?

Introduction

Milo and his friends follow clues and maps to solve a riddle. Maps are graphical representations of the Earth. There are several different types of maps, such as world maps, relief maps, topographical maps, and street maps, but they all have one thing in common. All maps share information that helps the user to determine position, so it is very important that they are accurate. The Earth is a spherical object, but most maps are flat representations. In this experiment, you are going to examine if flat representations of the Earth are the same as spherical representations.

Hypothesis

Is a flat representation the same as a spherical representation of the Earth?

Yes No

Materials

You will need:

- Blue balloon (with the continents drawn or printed on it)
- Flat map
- Pin

Procedure

1. Blow up your balloon so that it looks like a globe. (Note—If your balloon did not have the continents already printed on it, now is the time to draw them.)
2. Take your flat map and wrap it around the balloon, while attempting to match the continents up.
3. Record your observations.
4. Take the pin, pop your balloon.
5. Then, cut the balloon map between The Americas and Asia.
6. Try to stretch your balloon to match the map, and record your observations.

Observations and Results

Conclusion

Map Sketch

Label the following on the sketch below – Lines of longitude, Lines of latitude, Arctic Circle, Antarctic Circle, Prime Meridian, Equator, Tropic of Cancer, Tropic of Capricorn, North Pole, South Pole. Add the title: Lines on the Globe.

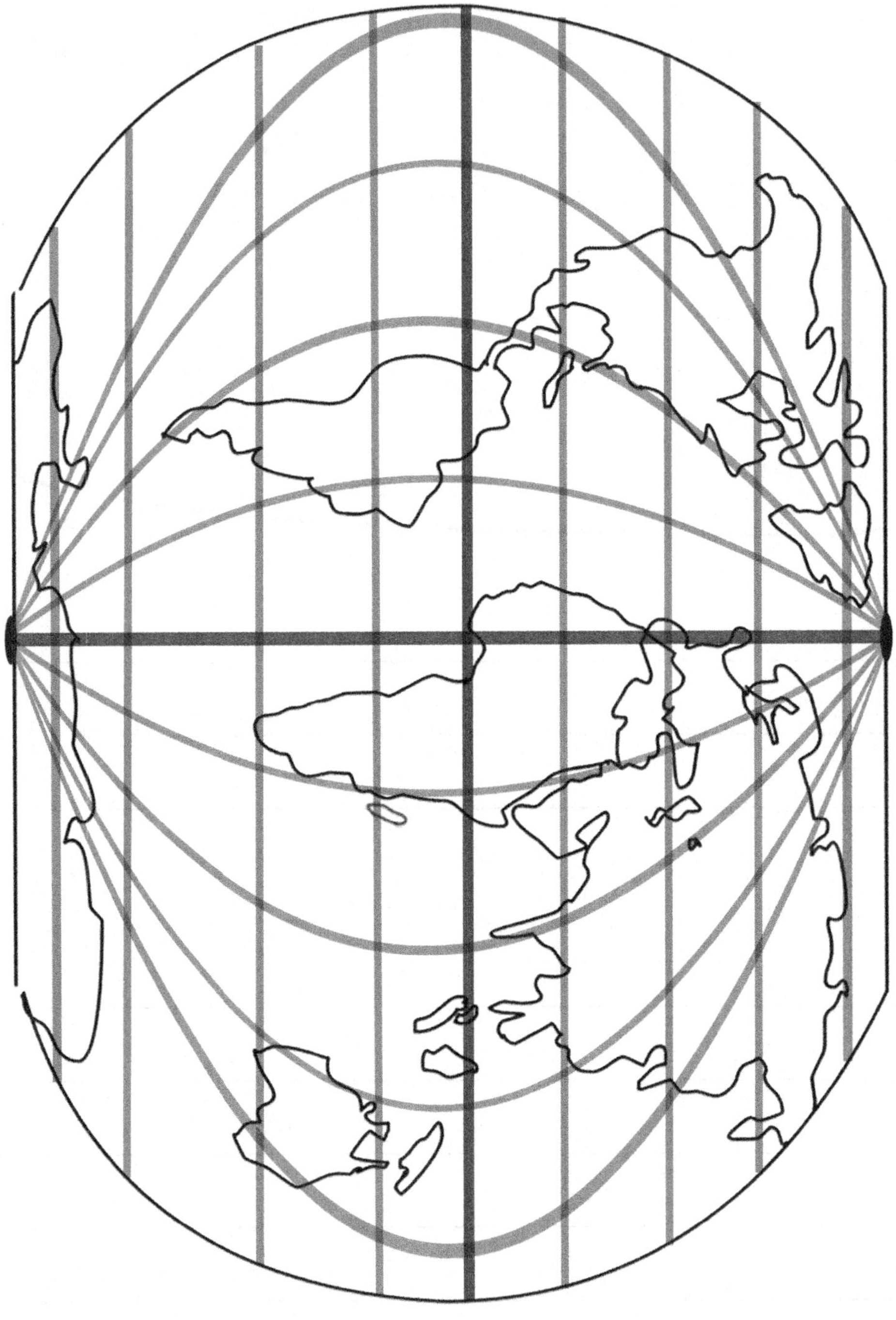

Map Notes

Your turn: Write what you learned about maps and mapping.

Need help getting started?
Answer the following questions: What is the purpose of a map? What is the purpose of the lines that you find on a globe? How are lines of latitude measured? How do lines of longitude differ from lines of latitude? How do cartographers represent a round Earth on a flat surface? How are satellites used in mapping?

Bonus Question: What would you include on a map so that a stranger could find your favorite place in your town?

Lesson 14: Rivers

Assignment Sheet

Read

- O Read Chapter 14 in *The PB Nanner Fluffer Test*: "Milo Caught a Ladybug."
- O Read the encyclopedia pages assigned by your teacher. Circle the ones your instructor wants you to do.
  - *Usborne Science Encyclopedia*: "Rivers" pp. 190-191
  - *Super Earth Encyclopedia:* "The Amazon River" pp. 124-125
  - *Everything You Need to Ace Science in One Big Fat Notebook:* "Freshwater Ecosystems" p. 505
  - (Other) ______________________________

Do

- O Do the "How is the course of a river determined?" experiment.
- O (Other) ______________________________

Write

- O Look up delta, estuary, and source. Add the definitions to the unit vocabulary sheet.
- O Label the "A River's Course" sketch.
- O Write a list of facts or a paragraph about rivers on the pages with the sketches.
- O (Other) ______________________________

____

Key Idea

A river is a moving body of water. Streams can flow together to form a river.

How is the course of a river determined?

Introduction

Milo and his friends went on a field trip to a river. A river is a natural flow of water from one area to a lake, sea, or ocean. Rivers typically contain fresh water and they are a very important part of the water cycle. Rivers are also home to a variety of wildlife and plant species. In this experiment, you are going to examine how a river's course is determined.

Hypothesis

I think that the course of a river is determind by ______________________________

____

____

____

Materials

You will need:

- Pitcher for water, Water, Cookie sheet, Paper cup, Straw, Dirt or sand, Small rocks, Books, Tape

Procedure

1. Poke a hole in the base of the cup, just large enough for the straw to fit tightly, using a pencil. Insert the straw and place the cup in the sink. Fill the cup halfway with water and watch what happens. If the water comes out of the straw like a stream, the cup is ready to use. If not, make the necessary adjustments.

2. Using several books, prop your cookie sheet up at an angle, so that water will flow down easily. Tape the cup to the top of the cookie sheet and fill it three quarters full with water. Observe what happens and record where the water went.

3. Dry the cookie sheet off and then sprinkle some dirt or sand all over the sheet. Refill the cup three quarters full with water, observe what happens and record where the water went.

4. Clean the cookie sheet off, sprinkle some dirt or sand all over the sheet, and

place a few rocks in various places. Refill the cup three quarters full with water, observe what happens and record where the water went.

Observations and Results

Stream	Observations
Plain Stream	
Stream with Dirt	
Stream with Dirt and Rocks	

Conclusion

Rivers Sketch

Label the following on the sketch below – source, upper stage, middle stage, tributary, lower stage, delta. Add the title: A River's Course

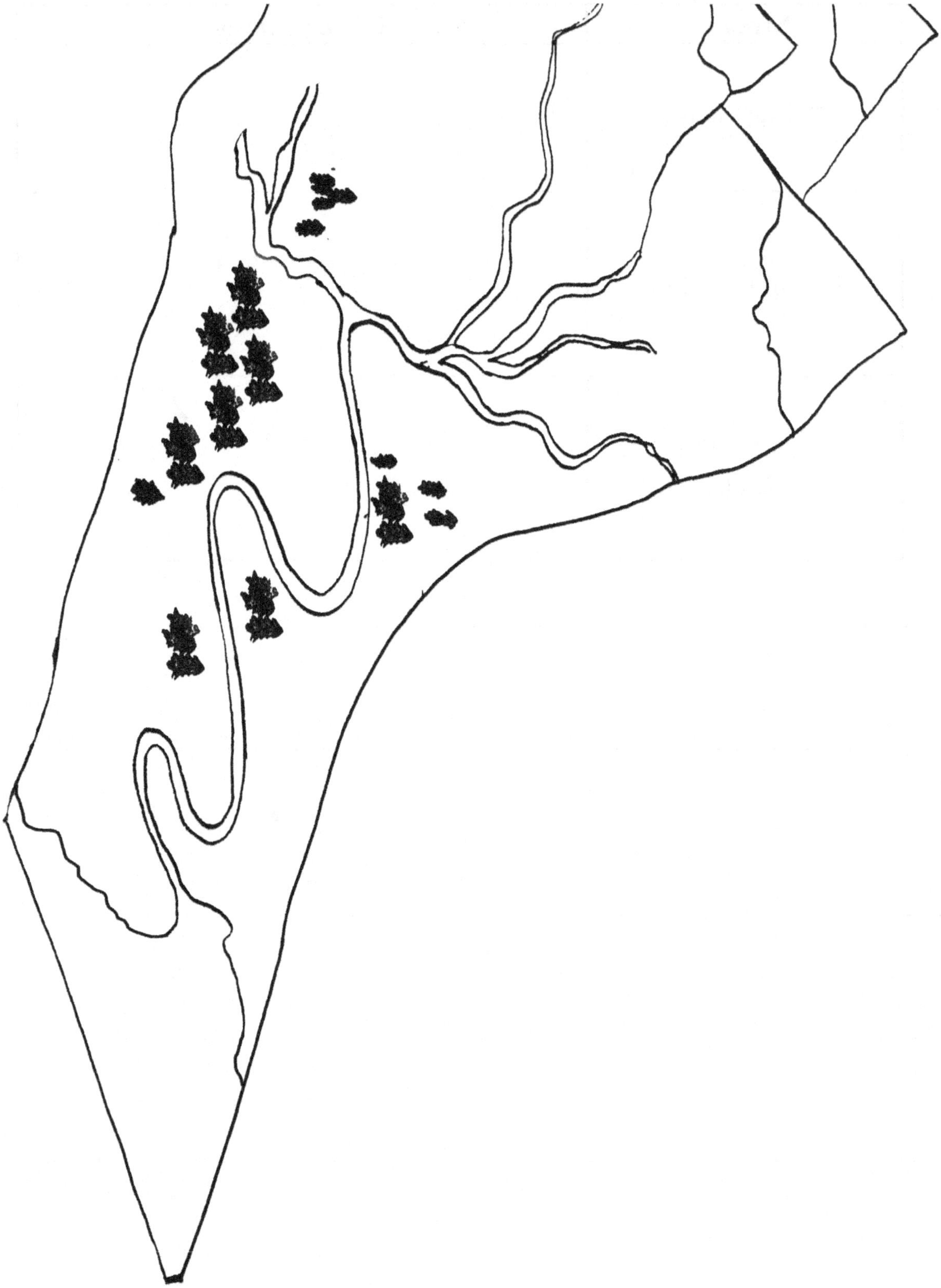

Rivers Notes

Your turn: Write what you learned about rivers.

Need help getting started?
Answer the following questions: Where does the water in a river come from? Describe a river's course. How do rivers erode the landscape around them? How is a load deposited by a river? Where is a river delta found and what happens at this point in the river?

Bonus Question: If a river were to disappear, what would happen to the surrounding environment?

Lesson 15: Oceans

Assignment Sheet

Read

- O Read Chapter 15 in *The PB Nanner Fluffer Test*: "Milo Wallowed Alone."
- O Read the encyclopedia pages assigned by your teacher. Circle the ones your instructor wants you to do.
    - *Usborne Science Encyclopedia*: "Seas and Oceans" pp. 188-189
    - *Kingfisher Science Encyclopedia*: "The Oceans" pp. 12-13, "The Ocean Floor" pp. 14-15
    - *Everything You Need to Ace Science in One Big Fat Notebook:* "Tides" pp. 205-206
    - (Other) ____

Do

- O Do the "Surface Versus Deep Currents" experiment. Fill out the experiment sheet as you do the experiment.
- O (Other) ____

Write

- O Look up coast, oceanic ridge, oceanic trench, and tides. Add the definitions to the unit vocabulary sheet.
- O Label the "The World's Major Seas and Oceans" sketch.
- O Write a list of facts or a paragraph about oceans on the pages with the sketches.
- O (Other) ____________________________________

____

Key Idea

The five oceans are the Arctic, Atlantic, Indian, Southern, and Pacific.

Surface Versus Deep Current

Introduction

Addison told Milo a story about her trip to the ocean, which is constantly in motion due to several currents that affect it. These currents are caused by several factors, such as wind, density, and temperature. Surface currents are caused by the wind blowing over the surface of the ocean, which then forms waves. Deep water currents are caused by cold water sinking and warm water rising. Deep water currents can also cause underwater waves. In this experiment, you are going to examine how these two different currents can affect the ocean floor.

Hypothesis

Can surface currents affect the ocean floor the same way that deep water

current do? Yes No

Materials

You will need:

- Aluminum bread pan or Plastic bin, Air dry clay, Water, Sand (1 cup), 2 Straws

Procedure

1. Have your teacher cut a small hole at the center of one end of your pan that is 1½ inches from the bottom. Insert your straw into the hole, so that when you blow through it the air skips over the bottom of the pan. Use some of the air dry clay to hold the straw in place and prevent water from coming out of the hole.
2. Next, roll out two lengths of air dry clay that are the exact width of the pan and place them 2 inches from each end of the pan. (See diagram for a visual reference.)
3. Fill your pan two-thirds of the way with water and add ½ cup of sand to the end of the pan that has the straw. Make sure that the sand stays between the air dry clay roll and the end of the pan.

4. Make some deep water currents by

blowing air through the straw at the end of the pan intermittently over one minute. Observe what happens and record your results.

5. Empty the pan, clean it out, refill it one third of the way with water and add ½ cup of sand in the same way you did in step # 4.
6. Make some surface currents by blowing air through the straw onto the surface of the water in the pan intermittently over one minute. Observe what happens and record your results.

Observations and Results

Currents	Observations
Surface Currents	
Deep Water Currents	

____

____

____

Conclusion

____

____

____

____

____

Oceans Sketch

Label the following on the sketch below – Arctic Ocean, Pacific Ocean, Atlantic Ocean, Indian Ocean, Southern Ocean, Mediterranean Sea, Red Sea, Black Sea, Caribbean Sea, Gulf of Mexico, Hudson Bay, Bering Sea, Tasman Sea, Coral Sea, Bay of Bengal, Arabian Sea, and North Sea. Add the title: The World's Major Seas and Oceans.

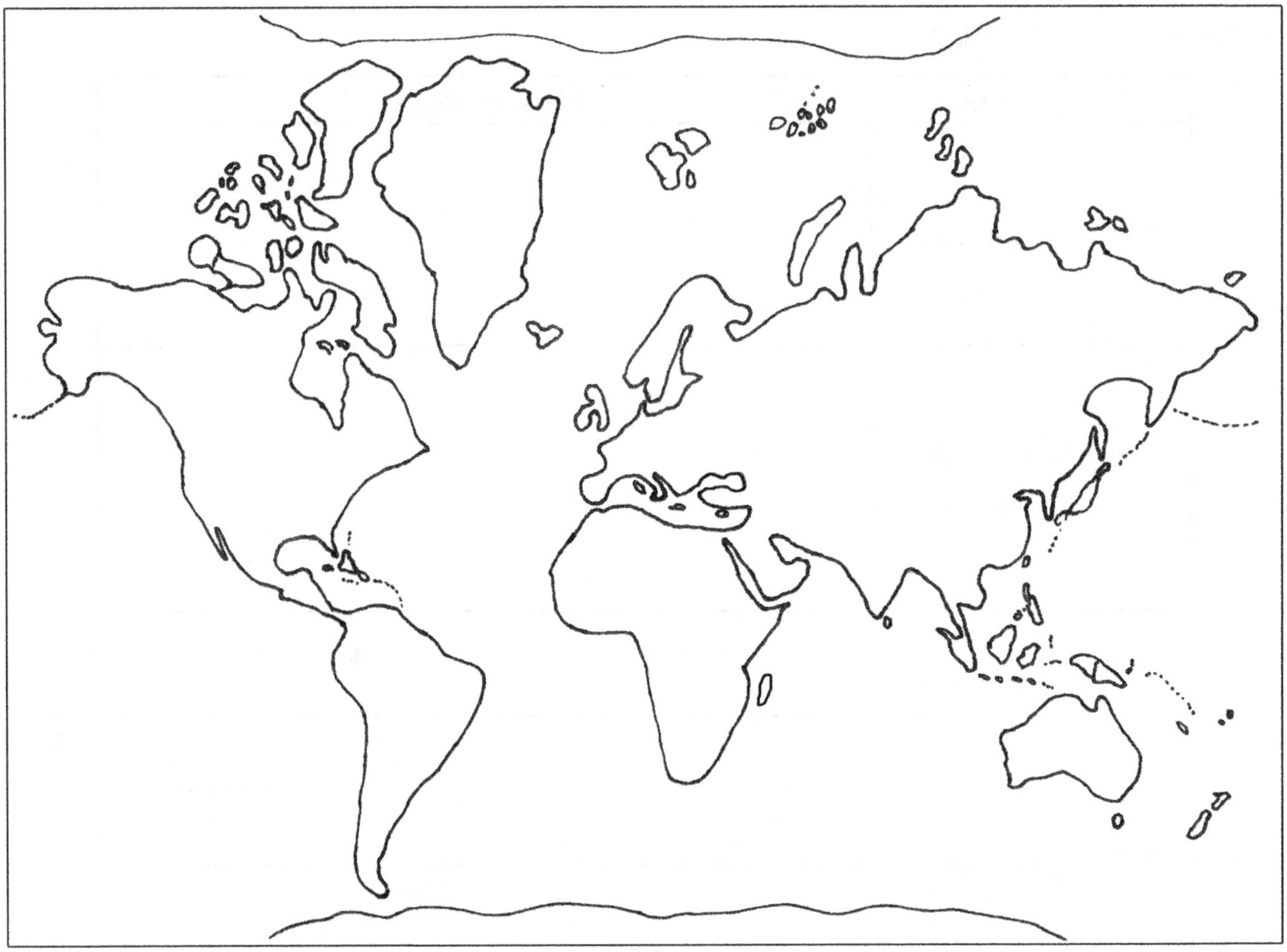

Oceans Notes

Your turn: Write what you learned about oceans.

Need help getting started?
Answer the following questions: How much of the Earth's surface is covered with water? How do currents work in the ocean? Why are the ocean depths so difficult to explore? Where do waves form and why do they crash on the shore?

Bonus Question: Would you like to live in the ocean? Why?

Lesson 16: Glaciers

Assignment Sheet

Read

- O Read Chapter 16 in *The PB Nanner Fluffer Test*: "Milo Enjoyed Friendsgiving."
- O Read the encyclopedia pages assigned by your teacher. Circle the ones your instructor wants you to do.
    - *Kingfisher Science Encyclopedia*: "Glaciers and Ice Sheets" pp. 34-35
    - *Super Earth Encyclopedia:* "Ice Caves: Mendenhall Glacier" pp. 106-107
    - *Everything You Need to Ace Science in One Big Fat Notebook:* "Erosion and Deposition" pp. 254-256
    - (Other) ______________________________

Do

- O Do the "What happens when glaciers melt?" experiment. Fill out the experiment sheet as you do the experiment.
- O (Other) ______________________________

Write

- O Look up deposition, erratics, glacier, and moraine. Add the definitions to the unit vocabulary sheet.
- O Label the "Anatomy of a Glacier" sketch.
- O Write a list of facts or a paragraph about glaciers on the pages with the sketches.
- O (Other) ______________________________

____

Key Idea

Glaciers are basically rivers of ice.

What happens when glaciers melt?

Introduction

Milo's Thanksgiving spread ended up looking like a morraine field left behind a glacier. Glaciers are huge masses of ice that flow downhill. Glaciers are solid, powerful forces that can cut through the landscape, picking up rocks and soil along the way. In this experiment, you are going to look at what happens when those glaciers begin to melt.

Hypothesis

I think that when glaciers melt ________________________________

____

____

Materials

You will need:

- Glacier Melt Model (Prepared the day before the experiment using cup, water, pebbles, and sand.)
- Large cutting board with a handle
- Large rubber band

Procedure

Day Before the Experiment: Make the Model

1. Start making your Glacier Melt Model by pouring a layer of pebbles and then a layer of sand into the bottom of a cup.
2. Then, cover with water and place in the freezer.
3. Once the water has frozen, repeat the process again until the cup has been filled, giving you at least three layers.

Day of the Experiment

4. Prop your cutting board up in the sink, handle-side up, so that the board is at about a 45 degree angle.

5. Remove the Glacier Melt Model from the cup using a little warm water. Use the rubber band to secure it with the pebble/sand side down to the top of the cutting board.
6. Observe what happens as the ice melts.
7. Record your observations and the time it takes for the model to melt.

Observations and Results

____

____

____

____

____

Conclusion

____

____

____

____

____

Glacier Sketch

Label the following on the sketch below – Cirque, Crevasses, Glacier's Snout, Meltwater, Movement of the Glacier (including arrows). Add the title: Anatomy of a Glacier.

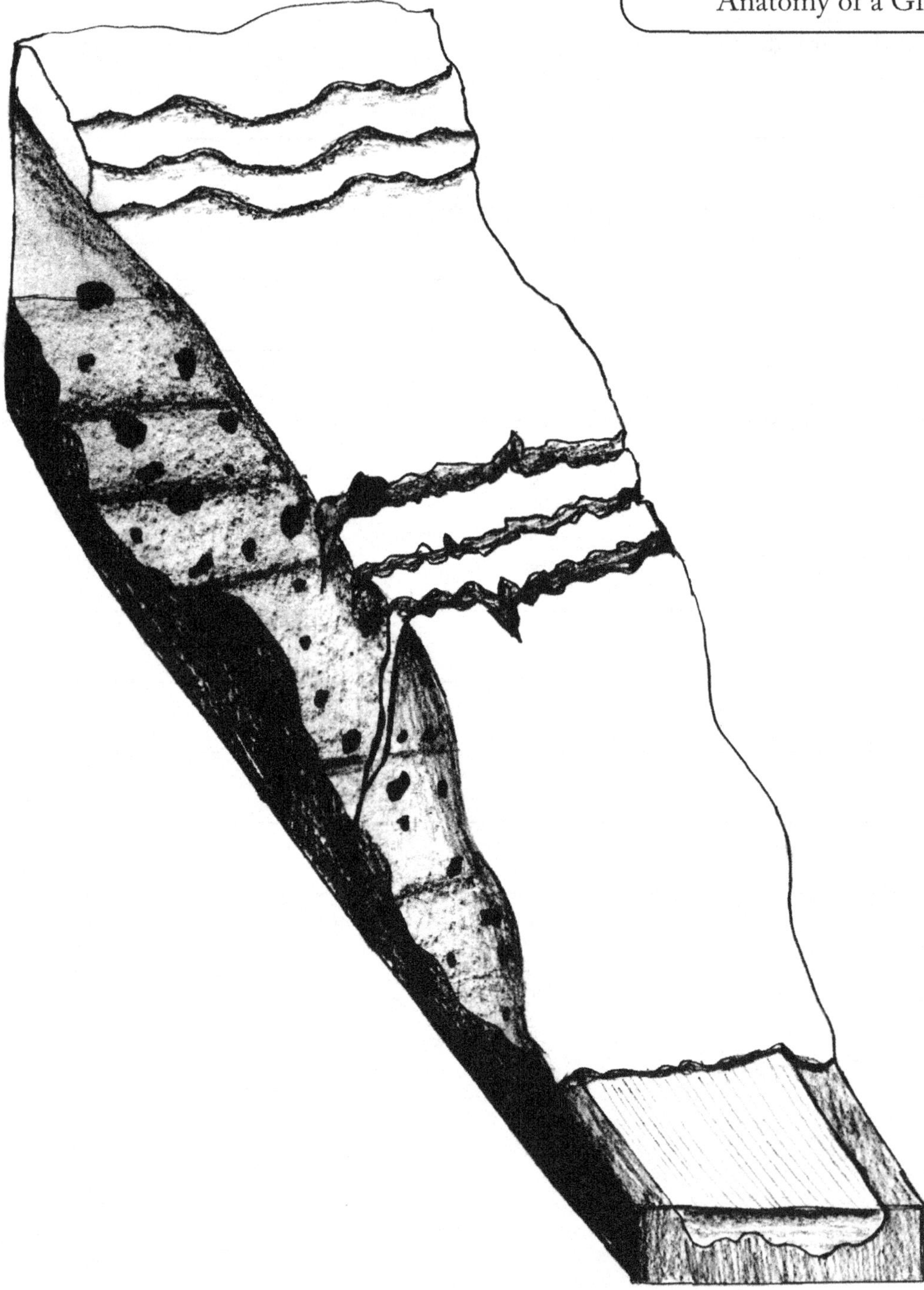

Glacier Notes

Your turn: Write what you learned about glaciers.

> **Need help getting started?**
> Answer the following questions: Where are glaciers found? How do glaciers form? What are some clues that let you know a glacier was once present in a valley?

Bonus Question: Why are glaciers blue?

Lesson 17: Biomes

Assignment Sheet

Read

- O Read Chapter 17 in *The PB Nanner Fluffer Test*: "Milo Executed the Plan."
- O Read the encyclopedia pages assigned by your teacher. Circle the ones your instructor wants you to do.
  - *Usborne Science Encyclopedia*: "Plants and People" pp. 290-291
  - *Kingfisher Science Encyclopedia*: "Biomes and Habitats" pp. 68-69
  - *Super Earth Encyclopedia:* "Living Earth" pp. 188-203
  - *Everything You Need to Ace Science in One Big Fat Notebook:* "Biomes" pp. 499-504
  - (Other) ______________________________

Do

- O Do the "How do desert plants live with so little water?" experiment. Fill out the experiment sheet as you do the experiment.
- O (Other) ______________________________

Write

- O Look up biome. Add the definitions to the unit vocabulary sheet.
- O Label the "Biome Around the World" sketch.
- O Write a list of facts or a paragraph about biomes and habitats on the pages with the sketches.
- O (Other) ______________________________

____

Key Idea

A biome is a larger region with a distinct climate.

How do desert plants live with so little water?

Introduction

Milo and his friends learned about habitats and biomes in co-op class. One of those biomes is the desert. The desert is a harsh environment in which only the hardiest plants and animals can survive. It is extremely hot during the day and very cold at night. The desert also receives very little rainfall. The plants and animals living there are well adapted to these conditions, like the camel, which can go for months without drinking water because its digestive system is extremely effective at extracting water. In this experiment, you are going to test how desert plants have adapted to survive with so little water.

Hypothesis

How do desert plants live with so little water? ______________________________

____

____

Materials

You will need:

- Paper towels, Cookie sheet, Wax paper, Rubber band, Water

Procedure

1. Moisten three paper towels with water until they are soaked, but not dripping, with water.
2. Lay one of the paper towels flat on the cookie sheet.
3. Roll the second paper towel up, secure it with the paper clips and lay it next to the flat paper towel on the cookie sheet.
4. Place the third paper towel flat on a piece of wax paper that is slightly larger, roll the two up together and secure with paper clips.
5. Then, lay the roll next to the other paper towels on the cookie sheet and set the sheet in a place where it will not be disturbed.

6. Check your paper towels the next day, make observations, and record the results.

Observations and Results

Sample	Observations the next day
Flat Paper Towel	
Rolled Paper Towel	
Rolled Paper Towel with Wax Paper	

Conclusion

Biomes Sketch

Label the following on the map below
– Choose one of the following biomes—arctic, desert, grasslands, rainforest, taiga, or tundra. Find where the particular biome is on the globe and label that region on the globe. Add the title: (Chosen) Biome

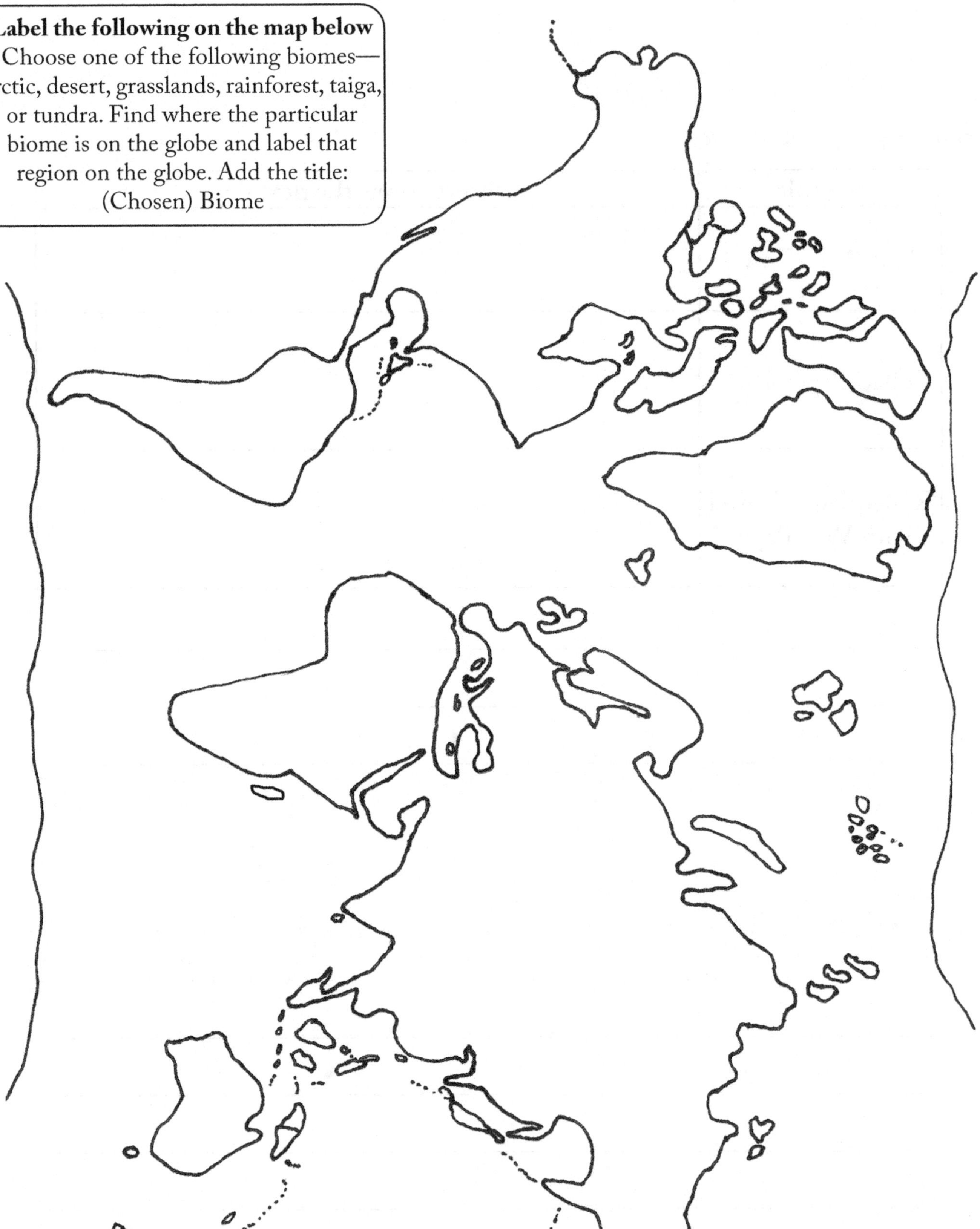

Biomes Notes

Your turn: Write what you learned about riomes and habitats.

Need help getting started?
Answer the following questions: What determines the boundaries between biomes? What is the difference between a habitat and a biome? What is a community?

Bonus Question: Which biome would you like to live in and why?

Lesson 18: Natural Cycles

Assignment Sheet

Read

- O Read Chapter 18 in *The PB Nanner Fluffer Test*: "Milo Celebrated the Holidays."
- O Read the encyclopedia pages assigned by your teacher. Circle the ones your instructor wants you to do.
    - *Usborne Science Encyclopedia*: "Natural Cycles" pp. 292-293, "Nutrient Cycles" pp. 334-335
    - *Usborne Illustrated Dictionary of Science:* "Carbon Cycle and Nitrogen Cycle" pp. 234-235
    - *Everything You Need to Ace Science in One Big Fat Notebook:* "Water Cycle, Nitrogen Cycle, and Carbon Cycle" pp. 491-494
    - (Other) ______________________________
- O Read the epilogue of *The PB Nanner Fluffer Test*: "Milo Has His Friends Over."

Do

- O Do the "Can I recreate the water cycle?" experiment. Fill out the experiment sheet as you do the experiment.
- O (Other) ______________________________

Write

- O Look up natural cycle and greenhouse gas. Add the definitions to the unit vocabulary sheet.
- O Label the "The Nitrogen Cycle" and "The Carbon Cycle" sketches.
- O Write a list of facts or a paragraph about natural cycles on the pages with the sketches.
- O (Other) ______________________________

____

Key Idea

Natural cycles show how nutrients are recycled in nature.

Can I recreate the water cycle?

Introduction

Milo helped his sister recreate the water cycle, which shows how water changes forms on the Earth, on a cookie.

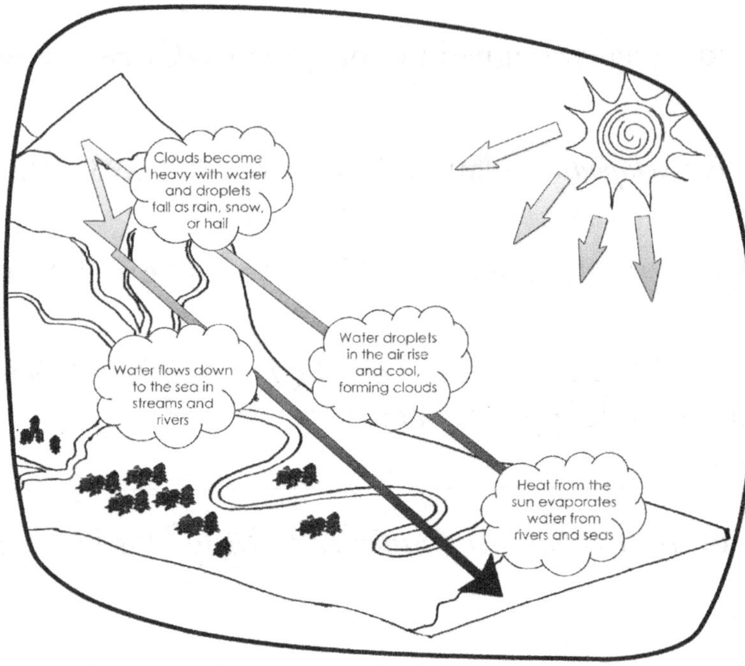

In today's experiment, you are going to try to create your very own mini-version of the water cycle.

Hypothesis

Can I recreate the water cycle? Yes No

Materials

You will need:

- Cup, Water, Plastic Baggie, Rubber Band

Procedure

1. Pour a little warm water into the bottom of your cup.
2. Cover it with the plastic baggie and use the rubber band to secure it in place.
3. Set the cup in a place where it will receive direct sunlight for the next two hours. (Note—If you do not have a sunny place to set your cup, set it under

a desk lamp instead.)

4. Check the glass after 2 hours.
5. Observe and record what has happened.

Observations and Results

Conclusion

Natural Cycles Sketch

Label the sketch with the following informaiton: Bacteria convert ammonia in the soil into nitrate, which is taken up by plants; plants also take up nitrogen from the air; plants are eaten by animals; as dead plants and animals decay, nitrogen is released into the soil. Add the title: The Nitrogen Cycle

Label the sketch with the following informaiton: When fossil fuels are burned, carbon dioxide is released into the air; plants take in carbon dioxide from the air to make food; animals take in carbon when they eat plants and release carbon dioxide when they breathe; as dead plants and animal decay, they release carbon dioxide into the air. Add the title: The Carbon Cycle

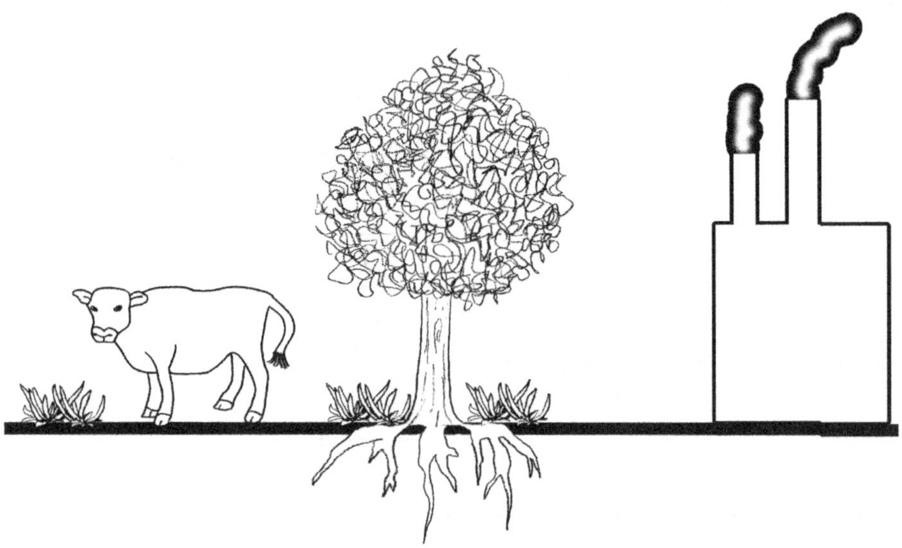

Natural Cycles Notes

Your turn: Write what you learned about natural cycles.

Need help getting started?

Answer the following questions: Briefly describe the nitrogen cycle. Briefly describe the carbon cycle. Briefly describe the water cycle. How are the natural cycles being upset?

SciQuest

Glossary

Level 100 Glossary

A

- Amphibian – A cold-blooded vertebrate that lives partly in water and partly on land.
- Autotroph – A living thing that makes its own food.

B

- Biome – A region of the Earth that contains unique plants and animals and is characterized by a distinct climate.

C

- Carnivore – An animal with specially shaped teeth that feeds mainly on meat.
- Cartographer – A person who makes maps.
- Cell – A tiny unit of living matter, the basic building block of all life.
- Classification – A way of identifying or grouping living things.
- Clutch – A set of eggs.
- Coast – The stretch of land that meets the sea.
- Crop – A pouch in a bird's digestive system where swallowed food is stored.

D

- Delta – A fan-shaped system of streams that is created when a river splits into many smaller branches before it enters the sea.
- Deposition – The dropping or leaving behind of rock and other debris by a glacier.

E

- Ectotherm – An animal whose temperature varies with its surroundings; ectothermic animals are also called cold-blooded.
- Endotherm – An animal with a constant temperature; endothermic animals are also called warm-blooded.
- Erratics – Large boulders that have been deposited away from their source by a glacier.
- Estuary – A wide channel that forms where a river joins the sea.

F

- Fledgling – A young bird at the time it leaves the nest.

G

- Gills – An organ used to breathe underwater.
- Gizzard – A chamber in an animal's stomach that grinds up food.
- Glacier – A mass of ice that gathers at the top of a land mass and slowly flows downhill.
- Greenhouse gas – Any gas that traps heat from the Sun, such as carbon dioxide.

H

- Hemisphere – One of the halves of the Earth created by an imaginary division along the equator.
- Herbivore – An animal that eats only plant food.
- Heterotroph – A living thing that eats other living things.

I

J

K

- Kingdom – The highest category into which living things are classified.

L

- Lines of Latitude (parallels) – Lines that run parallel around the globe, dividing the globe into flat slices; they never meet.
- Lines of Longitude (meridians) – Lines that run from the North to the South pole, dividing the globe into segments; they all meet at the poles.

M

- Mammals – A warm-blooded animal with hair that feeds its young on milk.
- Mammary gland – The milk-producing organ of a female mammal.
- Mantle – The mostly solid part of the Earth that lies between its crust and its core.
- Migration – A journey by an animal to a new habitat.

- Mitosis – The division of a cell nucleus to produce two identical cells.
- Monotreme – A mammal that lays eggs, such as a duck-billed platypus.
- Moraine – Rock, clay, sand, and other debris left by a glacier at the valley floor.

N

- Natural cycle – The exchanging of essential elements, such as nitrogen, carbon, and oxygen.

O

- Oceanic ridge – A raised ridge on the seabed caused by the movement of the Earth's crustal plates.
- Oceanic trench – A deep trench in the seabed formed when one place of the Earth's crust moves under another.
- Omnivore – An animal that eats both plant and animal food.

P

Q

R

- Reptile – A cold-blooded vertebrate with scaly skin.

S

- Scales – Small, overlapping plates that protect the skin.
- Source – The beginning of a river.
- Species – A group of living things than can breed together in the wild.
- Swim bladder – A gas-filled bag that helps a fish to float in water.

T

- Tadpole – The immature form of a frog or toad.
- Tides – The daily movements of the sea up and down the shore; they are caused by the gravitational pull of the moon.

U

V

- Vertebrate – An animal with a backbone.

W

X

Y

Z

Made in the USA
Monee, IL
13 September 2026

62980089R00079